ISBN 978-1-332-30735-7
PIBN 10311980

1 MONTH OF
FREE
READING

at

www.ForgottenBooks.com

By purchasing this book you are eligible for one month membership to ForgottenBooks.com, giving you unlimited access to our entire collection of over 700,000 titles via our web site and mobile apps.

To claim your free month visit:

www.forgottenbooks.com/free311980

English
Français
Deutsche
Italiano
Español
Português

www.forgottenbooks.com

Mythology Photography **Fiction**
Fishing Christianity **Art** Cooking
Essays Buddhism Freemasonry
Medicine **Biology** Music **Ancient
Egypt** Evolution Carpentry Physics
Dance Geology **Mathematics** Fitness
Shakespeare **Folklore** Yoga Marketing
Confidence Immortality Biographies
Poetry **Psychology** Witchcraft
Electronics Chemistry History **Law**
Accounting **Philosophy** Anthropology
Alchemy Drama Quantum Mechanics
Atheism Sexual Health **Ancient History**
Entrepreneurship Languages Sport
Paleontology Needlework Islam
Metaphysics Investment Archaeology
Parenting Statistics Criminology
Motivational

SILKWORMS.

BY

EDWARD A. BUTLER, B.A., B.Sc.,

Author of " Pond-Life : Insects," etc.

ARDVA·QVÆ·PVLCRA

LONDON :

SWAN SONNENSCHEIN, LOWREY, & CO.,

PATERNOSTER SQUARE.

—

1888.

CONTENTS.

CHAPTER I.

SILKWORMS.

CHAPTER I.

THE HISTORY OF SILK CULTURE.

"Patience and perseverance turn mulberry leaves into the silken robes of a queen." So runs an ancient Eastern proverb, which, if we strip it of its metaphorical signification, and regard it simply as a compendious method of stating what results when a certain series of natural processes is supplemented by the industry and artistic skill of human kind, may be appropriately placed at the head of our first chapter, as indicating the nature of the facts this little volume is intended to detail and illustrate. The fortunes of the silkworm and the mulberry tree are indissolubly associated, and when man steps in and patronizes the union between the two, there results an industry which has for ages been the support of millions of his race, and has supplied him with the most gorgeous of all those fabrics with which it has ever been his delight to adorn his person.

If we enquire who first kept silkworms, and whether they were kept for pleasure or for profit, we shall find that, while it is easy enough to give a traditional reply which has all the sanction of a very hoary antiquity, it is not by any means so easy to say how much reliance may be placed on this, and how far it represents actual history; for the silkworm is now so entirely a domesticated animal, that, like the dog and some other of man's dumb friends, it is not met with in the wild state, at least in the form in which it is reared, and but for man's care would, in the course of a twelvemonth, disappear from the face of the earth. So long has it been a companion of man, that the history of the first reclamation from the wild state of the "dog of insects," as it has been termed by one writer, is mixed up with myth and fable, and well-nigh lost in the mists of antiquity. But, notwithstanding this,

we may say with certainty that the progenitors of all the silkworms
now under man's care came originally from the East; and all
traditions point to China as the country where the insect was first
domesticated, and whence it has spread, till its cultivation now
employs thousands upon thousands of persons, men, women,
and children, in very varied and widely distant parts of
Asia, Europe, and America. That it should have attracted
attention at a very early age, in those countries in which it was
indigenous, is not at all to be wondered at, for the beautiful contrast of the silken cocoons entangled amongst the dark foliage of
the mulberry trees, could hardly fail to appeal to the sensibilities
of the inhabitants of those districts, even if, as may have been the
case, the cocoons were not at that time quite so beautiful as they
are now. But it is scarcely likely that the mere appearance of
these "things of beauty" would convey any suggestion of the
possibility of utilising them in human economy, at least until men
had become familiarised with the use of textile fabrics composed
of other materials.

As China seems to have been the cradle of the silk manufacture,
and the native country of the silkworm, so it is to the Chinese
that we must look for the earliest accounts of the insect and its
cultivation; and we find, accordingly, that that ancient people
possess a tradition which professes to detail the discovery of the
material and the origin of the manufacture. They have an
ancient book called the "Silkworm Classic," which contains an
account of a certain empress, the wife of the celebrated Hwang-te,
who is said to have flourished about 2,600 years B.C. This
lady, whose name was Se-ling-she, is said to have been the first to
undertake the trouble of rearing silkworms, and the account proceeds to state that she was induced to take this step by her husband, who was anxious that, after all the schemes he had himself
undertaken for the benefit of his people, his wife also should do
something to prove herself a benefactress to her race, so that
their names might together be handed down with glorification to
a grateful posterity. Accordingly she undertook the task, and in
concert with her husband, who apparently had beforehand some
suspicion of the nature of the advantage that might be reaped,
was led thereby to the discovery of the uses to which the silk
might be put, and to the consequent invention of garments composed of the newly-discovered material.

If any reliance could be placed on this date, then the silkworm
would by this time have been bred by mankind generation after
generation, for a period of some 4,500 years. But whether this
be the exact date or not, certain it is that its cultivation has been

practised in China from a very remote antiquity, and that the greatest possible encouragements have been given to the art by the successive sovereigns, and that a vast number of laws and regulations were enacted with a view to the advancement and protection of the industry.

The empresses, in particular, in commemoration of the inestimable discovery of their ancestress, took the rearing of silkworms under their own special patronage, to which end there was an enclosure attached to the palace for the cultivation of mulberry trees, and just as the emperor showed his desire for the advancement of agriculture by going in state each year to turn the first clods with the plough, so his wife, surrounded by the ladies of the court, went annually in state into the royal mulberry garden to pluck with her own hands the leaves of three branches which the ladies-in-waiting lowered till they were within reach, and thus she inaugurated the silkworm season.

A somewhat different account is, however, given in one Chinese work of great antiquity, which seems to imply that the emperors also took an active interest in the occupation, their wives appearing in a more subordinate capacity. We read that, in ancient times, the emperor and his princes had a public mulberry garden, and a silkworm establishment erected near some river. "On the first day of the third month of spring," the sovereign, who evidently meant business, came out dressed in "a leather cap and plain garments," and, apparently finding some difficulty in deciding otherwise, ascertained by lot which was the chief of his three queens ; the selection having been made, the fortunate lady was told off, together with the most honourable of the other members of his harem, to attend to the rearing of silkworms in the above-mentioned establishment. These then, having received their commission, brought the eggs and washed them in the river that ran close by, after which they picked the mulberry leaves, aired and dried them, apparently by placing them in their bosoms, and proceeded to feed the worms. At the close of the season a special ceremony, though of an exceedingly simple character, was gone through. The ladies brought the cocoons to their prince, and he presented them again to his chief queen ; whereupon she said, " This is the material of which your highness's robes are to be formed." Having said this, she covered herself with her robe, and received the cocoons. The ladies of the court, on such occasions, received presents of sheep, apparently in order to reward them for their zeal in silkworm culture, and to serve as an inducement to them to undertake the like duties the next season.

As examples of the laws made to regulate and encourage the industry, as well as illustrations of its antiquity in China, the following may be cited. On one occasion, about ten centuries B.C., an order was issued by the officer who adjusted the price of horses, forbidding the people to rear a second brood of silkworms in one season. And lest it should not be quite clear to the ordinary intellect, what connection there could be between horses and silkworms, to justify such an edict emanating from such a source, we are kindly informed that horses and silkworms were considered by the astrologers to belong to the same constellation, and that therefore they must be of the same origin (this was Evolution with a vengeance!), and that, as it was unlikely that two things of the same nature should prosper at the same time, the rearing of a second brood of silkworms must be forbidden, lest it should be of some disadvantage to the horses !

Again, coming down to later times, though still before the Christian era, we read as follows : " In the first month of spring, orders were issued to the forester not to cut down the mulberry trees ; and when the cooing doves were observed fluttering with their wings, and the crested jays alighting upon their mulberry trees, the people were to prepare their trays, frames, etc., for the purpose of rearing the silkworms." What a pity it is that the legal English of to-day is not as high-flown and poetical as the Chinese of two thousand years ago ! The silkworm season seems to have been a kind of Lent, for again we read : " In the spring season, when the empress and her ladies had fasted, they proceeded to the east, and personally engaged in picking mulberry leaves ; on this occasion the married and single ladies were forbidden to wear their ornaments, and the usual employments of females were lessened, in order to encourage their attention to the silkworms. When the rearing of the silkworms was completed, the cocoons were divided and the silk weighed, each person being rewarded according to her labour, in order to provide dresses for the celestial and ancestorial sacrifices : in all this none dared indulge in indolence."

Many other peculiar arrangements might be cited, but these are sufficient to show with what care this important industry was fostered in the Celestial Empire. But at the same time, with the exclusiveness which has always characterised this peculiar people, the Chinese adopted the strictest possible precautions to prevent the rest of the world from profiting by their valuable discovery, and to keep the monopoly of the manufacture ; and to so great an extent was this carried that it was death to export from China silkworms' eggs, or to supply any such information as would enable foreigners to prepare silk for themselves. And thus it happened that, while at a very early period the manufacture spread so widely in China itself that silken garments were worn even by the peasantry, the very material remained to all but the nations immediately around, an unknown article. But at some early age, one knows not when nor how, India also learnt the secret, and hence after a while a considerable trade in the manufactured article was carried on amongst the Asiatic nations generally.

There is a pretty little legend as to the introduction of silk culture into Japan, which, though entirely mythical, is worth relating. There was a certain king of India who, following the fashion of the times, had many wives and a very large family ; his favourite wife, however, was, much to his disappointment, for a long time childless. At last a child was born to them, but when the father heard that it was a daughter, instead of the son he had so passionately longed for, his exasperation was unbounded, and he expressed himself in no very guarded terms on the subject. His words were by some busybody reported to the mother of the unfortunate infant, and she was so deeply affected by them, that she soon afterwards died, leaving her husband a prey to the bitterest remorse. So inconsolable, indeed was he, that a few weeks after, he followed his wife to the grave, leaving the baby, Youan Thsan (*i.e.*, silkworm) by name, to the tender mercies of its very numerous step-mothers. Now step-mothers are not unfrequently very undesirable persons to be entrusted with the well-being of a helpless orphan, and the present assemblage of ladies was no exception to the rule. They argued, rightly or wrongly, that this "latest arrival" would considerably interfere with the ambitious projects they had been forming for their own progeny, and that therefore it would really be an excellent thing if the little maiden could be comfortably and quietly disposed of. Having once made up their minds to this course, they did not take long to carry out their intention, but, in the attempt, they met with unexpected difficulties, for the young lady proved a good deal more self-assertive than they had calculated for. That their own hands might not be stained with its blood, they had the child conveyed to a desert infested with lions, and there abandoned. Presently a young lion approached, and assumed a threatening attitude. But the royal infant, putting on great dignity and boldness of manner, accosted the enemy with the question, "Who are you?" The reply came in majestic tones, "My father is the king of beasts." "But my father," responded the child haughtily, "was the king of men!" Thus rebuffed, the beast skulked away, and went off to summon its elders to its assistance. The infant profited by its retreat and returned to the palace. The stepmothers were greatly surprised at the failure of their little plan, and had the child once more conveyed away, this time to a valley which was much frequented by eagles. An eagle soon arrived on the scene and carried the child to its nest, and then departed, seeking other prey. In the nest a colloquy took place between the princess and the eaglets, similar to what had occurred with the lion's whelp, the reply to the child's question being, of course, "Our father is the king of birds," and so on as before. Again the child returned to the palace. This time the stepmothers put her on a desert island. Here she was found by a fisherman, who at first claimed her as his property, as an ocean waif. But the little princess, still keeping up that dignified bearing, which had already stood her in such good stead, informed her captor that her father was "king of kings and of men," whereupon he treated her kindly and took her back to the palace again.

This was too much for the stepmothers, who began to fear that they should always have the child on their hands, and they therefore decided to adopt more energetic measures. They gave orders that she should be buried alive in the courtyard of the palace. However, the sexton who performed the burial was compassionate, and threw the sods in so lightly that the child was able to breathe. That night an earthquake took place, and once more the irrepressible princess appeared above ground un-harmed. But the ingenuity of her enemies was not yet exhausted, nor their ill-will quenched. They therefore had the trunk of a mulberry tree hollowed out, and putting the infant inside, they sent the whole apparatus adrift on the ocean. Tossed about by wind and wave, the little bark with its precious freight, was at last stranded on the coast of Japan, on reaching which the poor little princess, utterly exhausted by the perils to which she had been exposed, expired. But her sufferings touched the hearts of the powers above, and the poor little body was transformed into a silkworm, which fed on the mulberry tree. The tree took root, flourished, and supported generation after generation of silkworms descended from the worm-princess. Thus was the silkworm introduced into Japan, and to this day, the four successive moults to which we shall have occasion to refer in the next chapter, are called the time of the lion, of the eagle, of the canoe, and of the courtyard.

This is the poetry and romance of the subject; the simple prose and history of the matter seems to be that eggs and trees were conveyed from China to Japan about the middle of the fifth century A.D.

The rearing of the silkworm became very popular in the East, and was speedily adopted as one of the chief occupations of the Chinese, by whom it was so highly esteemed that they deified the traditional foundress of the art, Se-ling-she, and called her the " Goddess of the Silkworm." Skill in silk-weaving, also, was regarded as a gift from heaven, as the following Japanese story testifies. There was once a youth named Toung Young, who was not in very flourishing circumstances. When his mother died, the expenses of the funeral appear to have fallen upon him, though the father was still alive, and the bereaved young man, with that filial piety which is esteemed the highest of the virtues among the Japanese, spent his all in procuring a coffin. Before he had time to recover from this drain on his exchequer, his father also died ; but the devoted son, determined not to be deprived of the credit of showing the honour due to the departed, hesitated not to sell himself, having nothing else on which to raise money, to a fellow-citizen, in order that he might procure the means to give his father a respectable funeral. Having accomplished this, he was on his way to fulfil his bargain with his purchaser, when he was met by a girl of remarkable beauty, who, to his unutterable surprise, and somewhat to his discomposure, offered to share his fortunes. He felt that his circumstances were scarcely such as to justify him in accepting such an offer, much as he may have desired to do so. He therefore explained to the lady the straits in which he was placed ; but she was not to be deterred from her purpose, and she volunteered to go with him to his employer and give her services in weaving. This turned out an excellent arrangement, for within a month she wove one hundred pieces of silk of patterns never seen before, but of marvellous beauty ; these she offered as a ransom for Toung Young, and needless to say the offer was accepted. The young man thus freed from his contract, started joyfully homewards with the engaging maiden who had done so much for him, dreaming of the domestic bliss that he hoped was in store for him. But on reaching the spot where she had first met him, the young lady coolly bade him farewell, and in explanation of her conduct, revealed to him that she was a heavenly messenger sent to reward his filial devotion, and that, having performed her mission, she had but to return whence she came. She thereupon ascended to the sky. The poor young man, thus rudely awakened from dreamland, returned to his employer and resumed the occupation of silk-weaving. But by carefully imitating the marvellous designs

worked by the angelic messenger, he succeeded in producing new
goods which were so greatly superior to those that had hitherto
been known that his fortune was speedily assured, and, of course,
" he lived happy ever afterwards."

For many hundreds of years the silk trade was confined to Asia ;
Europe did not become acquainted even with the manufactured
article, far less with the method of its production, until a comparatively late date.   Aristotle, who flourished about 360 B.C.,
is the earliest European writer from whom we learn anything
about it.   In his time the south-east of Europe was supplied with
silk fabrics from the island of Cos, off the west coast of Asia
Minor, and hence we may gather that the manufacture had
travelled thus far westward at least as early as 400 B.C.   However,
the silk was in all probability not produced there ; but sub-
stantial silk fabrics were imported from China into the island,
where they were unwoven and respun in a much finer condition,
the compensation for the trouble of the operation being the increase in quantity arising from the greater fineness of the thread.
The method of accomplishing this is said to have been invented
by a woman named Pamphila.

The manufactory at Cos became highly celebrated, and the
fabrics made were remarkable not only for their extreme costli-
ness, but also for their extraordinary thinness and delicacy ; they
were so thin as to be almost transparent, so that when a person
was clothed in silk, it served but little as a protection or as a con-
cealment to the body.   Such garments, called from the place of
their manufacture Coan vestments, were worth their weight in
gold, and none but the richest could indulge in so great a luxury.
The wealthiest Romans in the most luxurious times of the Empire,
spent vast sums of money on such adornments, the rage for silken
clothing not being confined to the ladies, but extending also to
the men of more effeminate type, though its use as a masculine
adornment was sternly condemned by the more sturdy Romans
of the period.

But though they were familiar with the material they had the
most erroneous ideas as to its nature.   It was a natural product,
indeed, but how it was produced or even whether it belonged to
the animal or to the vegetable kingdom, they were not by any
means agreed.   It had some connection with insects, and some
with trees, but when the attempt was made to assign to each its
proper share in the production they got hopelessly at sea.   Their
enquiries about it had shown them that its original source was
somewhat farther east than they had before supposed, and that
it came from a people called the Seres, but who these were, or

where they lived, they did not know; and indeed the name simply means the "silkworm people," the Greek name for silkworm being " *ser.*" Apparently under the name Seres the Chinese were referred to, the name for the insect among that people being *see*, from which, no doubt, the Greek name was taken; and the prevalent idea was that silk was in some way or other yielded by the woods in the districts inhabited by that people.

Pliny, the Roman naturalist, who flourished during the earlier part of the first century A.D., gives in his account of the silkworm a curious mixture of zoological and botanical rubbish, which we will quote in the quaint words of an old translator, as being, in consideration of the extreme antiquity of the notions expressed, a more appropriate clothing for the Latin than a modern English rendering would be. The passage runs thus :—" They build their nests of earth or clay, close sticking to some stone or rock, in manner of salt; and withall so hard, that scarcely a man may enter them with the point of a spear. In which they make also wax, but in more plenty than bees; and after that, bring forth a greater worme than all the rest before rehearsed. These flies engender also after another sort, namely, of a greater worme or grub, putting forth two hornes after that kind: and these be certain canker-wormes. Then these grow afterwards to be Bombilii, and so forward to Necydali: of which in six months after come the silke-wormes Bombyces. . . . It is commonly said, that in the Isle Cos there be certain silkwormes engendered of flowers; which by the meanes of rain-showers, are beaten downe and fall from the Cypres tree, Terebinth, Oke and Ash; and they soon after doe quicken and take life by the vapor arising out of the earth. And men say, that in the beginning they are like unto little Butterflies naked, but after a while (being impatient of the cold) are overgrowne with haire; and against the winter, arme themselves with good thick clothes; for being rough-footed, as they are, they gather all the cotton downe of the leaves which they can come by, for to make their fleece. After this, they fal to beat, to felt and thicken it close with their feet, then to card it with their nailes; which done they draw it out at length, and hang it betweene branches of trees, and so kembe it in the end to make it thin and subtill. When al is brought to this passe, they enwrap and enfold themselves (as it were) in a round bal and clew of thread, and so nestle within it. Then are they taken up by men, put in earthen pots, kept there warme, and nourished with bran, untill such time as they have wings according to their kind; and being thus well-clad and appointed, they be let go to do other businesse."

Such was the outlandish explanation given to the western world eighteen centuries ago, of the origin of that beautiful fabric which had only recently become known in Roman society, and was so much prized for its softness and brilliance, and for the readiness with which it took the most splendid dyes.   It is plain that, in this account, the author has mixed up several totally distinct insects, but as he had never seen them himself, and spoke only from hearsay, his mistakes are the more pardonable.   What his Bombilii and Necydali were, he probably knew no better than we do, for he simply borrowed here terms which Aristotle had used in the same connection four hundred years before.   Moreover, the Greek philosopher, though not giving utterance to statements quite so ridiculous as his Latin successor, was not very greatly his superior in the correctness of his information on this particular subject.   Another writer gravely states that silk consisted of the entrails of a spider, which was fed for four years on a kind of paste, and after that on willow leaves until it actually burst with fat !

So great was the demand for silken garments in South Europe in the first two centuries of our era, that it outstripped the supply of imported material, which therefore became rarer and more expensive, till at last, by the middle of the third century, we find the Emperor Aurelian actually forbidding his wife to have a silken robe, in consequence of its enormous expense.   But shortly after this, it became commoner, though frequently of an inferior quality, and fabrics of mixed woollen or linen and silk were often worn. By the middle of the sixth century a considerable trade in silk had sprung up, and Persia was the country through which the products of Eastern Asia were introduced to the Europeans, because the great caravan routes from the East had their termini in her territory.   Persia therefore took every precaution against letting the silk trade escape from her hands, and in consequence she was able to keep up the prices to an inordinate height, and made a very good thing of the business, much to the annoyance of the Europeans, who resented the passing of so much wealth away from themselves.   There seemed, moreover, no reason why silk culture should not be practised in Europe, since the mulberry tree grew wild in the south-eastern parts.   But all efforts to procure silkworms for Europe, in the course of fair and open trade, proved unavailing, so jealously was the industry guarded by the Asiatics.   What, however, could not be accomplished in an open and straightforward manner, was at last effected by stratagem.

In the reign of the Emperor Justinian, about the middle of the sixth century A.D., two Persian monks of the order of St. Basil, who had gone as missionaries to India, and had thence penetrated

to China, conceived the idea, regardless of the interests of their
native country, of watching carefully all the processes of silk
culture, and then endeavouring to obtain a supply of silkworms'
eggs, and, if possible, transporting them to Europe. Stimulated
by the promise of great rewards on the part of the emperor—a
somewhat necessary provision, for they knew they were running
great risks to life and liberty—and actuated, so it is said, also by
a feeling of resentment that so valuable an industry should remain
the exclusive property of "unbelievers," they kept their eyes open,
of course saying nothing to any one, and, after a time, thoroughly
mastered all the details of silk culture. As winter approached,
they managed to obtain a considerable supply of eggs, and well
knowing that if they were caught conveying them out of the
country they would have to pay the penalty of their lives, they
carefully concealed them in the hollow canes which they carried
as pilgrims' staves. In this way, they conveyed their precious
property throughout the long and tedious journey to Constantinople, where at length they safely arrived.

They at once communicated the success of their attempt to the
emperor, who was, of course, delighted, and undertook to establish
the art of silkworm rearing amongst his people, promising them
every encouragement, if they would give the necessary instructions.
So the experiment was commenced. The eggs were hatched,
and the young grubs fed on the wild mulberries of the district.
They flourished, produced silk, and laid the foundations of a new
generation for the following season. This was the beginning of
silk culture in Europe, and from this insignificant commencement
there sprang the whole of the important industry of silkworm
rearing, which, continued from year to year and from generation
to generation, has grown to such dimensions that it now gives
employment to an enormous number of persons in Italy, France,
Spain, Turkey, Greece, and South Russia, and produces an annual
supply of raw silk weighing no less than three thousand five
hundred tons—representing at 15s. a pound, a value in hard cash
of £5,880,000.

Thus by the enterprise of two monks, an important blow was
struck at the Persian trade, and a valuable addition made to the
industrial occupations of Europe; but the Emperor Justinian,
having had so much difficulty in starting the new industry,
mined to maintain a firm hold upon it, and therefore he
confined it to his own district, so that for centuries Rome
and Constantinople that supplied Europe with the silk
required. About the beginning of the eighth century
was introduced into Spain by the Arabs, b

twelfth century that, partly through the influence of the Crusades, it spread to other parts of Europe. By Roger IL, king of the two Sicilies, some Greek weavers were, in 1130, forcibly carried to Palermo, together with silkworms' eggs and mulberry trees, and there they were compelled to carry on the manufacture for the benefit of Sicily. Hence it gradually spread northward throughout Italy, which at the present day is the chief silk-producing country in Europe, and indeed, is second only to China throughout the world. Florence early became a centre of the silk trade, and in order to ensure success, each peasant in Tuscany was compelled to plant at least five mulberry trees on the land he cultivated. Venice also took a prominent position in the new industry, and at the commencement of the fourteenth century, three thousand people were there employed in it.

Towards the close of the sixteenth century, silkworms were introduced into France, by the exertions of the king, Henry IV. Mulberry trees had for a long time been grown in France, the first ones having been introduced from Italy nearly two hundred and fifty years before, and the king argued that as his country produced mulberry trees, it might just as well be provided with silkworms to feed upon them. He was much assisted in his scheme by a celebrated agriculturist of the time, Olivier de Serres, who was the first Frenchman to publish instructions as to the growing of mulberry trees and the rearing of silkworms. At the suggestion of this man, the king imported from Italy twenty thousand young mulberry trees, and large quantities of silkworms' eggs, and these he caused to be distributed in those parts of France that seemed most suitable for the undertaking. The experiment succeeded, and the industry was fairly established. In consequence, however, of the many political troubles of that unfortunate country, the manufacture has had many ups and downs, and in addition to this, the ravages of disease amongst the silkworms, a quarter of a century ago, almost ruined it. It has, however, maintained its ground notwithstanding all obstacles, and still flourishes, especially in the south, where mulberry trees are grown, and the insect reared on very extensive scale. Lyons is the centre of the industry.

In our own country we are now so familiar with silk that it is extremely difficult to realise how modern an article of luxury it is, that less than three hundred years ago it was so rare a subject that even royalty was compelled, on an emergency, to a simple article of dress made of that material, for none of could be found in the royal wardrobe. James I. when Scotland, wishing to appear with all due pomp before so personage as the English ambassador, was greatly

But though this first attempt fell through, others since carried out at different places, as Slough, Nottingham, Yately, etc., have been more successful, and have demonstrated that it is quite possible for good silk to be produced in this country. The climate, however, is too changeable to permit of the permanent establishment of silk-production, and the expenses attending the process have always been found to be too heavy for it to be financially a success ; it has been found cheaper to import the raw material than to produce it on the spot. The rearing of silkworms in our own country, therefore, is still merely a pastime, and " educations " as the French call them, are carried out only on an extremely insignificant scale, far smaller than any one would dream of undertaking if it were done with a view to utilising the silk.

Not only is this the case, but until the year 1718, England had to depend upon foreigners even for silk thread; she did not practise the art of reeling and " throwing " the silk from the cocoons, and so preparing it for manufacturing purposes. But in that year Lombe, of Derby, having gone to Italy disguised as a common workman, and taken drawings of the silk-throwing machines in use there, set up, on his return, a mill in his native town. This was on a very large scale, and produced daily no less than 318 millions of yards of prepared silk thread.

We must at the present day, therefore, distinguish three totally distinct branches of the silk industry, the production of the raw material, the transformation of this into thread suitable for manufacturing purposes, and lastly the weaving of this into the various beautiful fabrics used so largely for clothing and other purposes, and it by no means follows that the same country will undertake all three processes.

CHAPTER II.

THE SILKWORM—ITS FORM AND LIFE HISTORY.

HAVING thus given a brief outline of the history of silk culture, we may now proceed to enquire what is the nature of the animal itself to which mankind is so greatly indebted, which has contributed so largely to his comfort and delight, has during so many ages proved the support of so many millions of his race, and has even brought itself into prominence as a cause of international jealousies. We are most familiar with it as a pale, worm-like creature of lethargic habits, and much given to gormandising ; this, perturbed because his hose seemed scarcely to match the rest of his costume, and he greatly longed to see his legs encased in the splendour of silken hose; but as he had none of his own, he was reduced to the ignominy of begging the loan of a pair from the Earl of Mar, at the same time pathetically appealing to the loyalty of this courtier with the words, "Ye would not, sure, that your king should appear as a scrub before strangers." Queen Elizabeth too, it is well known, was particularly pleased with the present of a pair of silk stockings, and with the characteristic Tudor love of dress, found so much pleasure in wearing them that she persistently refused ever after to wear any other kind.

The manufacture of silk was introduced into this country in the reign of Henry VI., but whatever progress it made was due chiefly to acts of religious intolerance on the Continent. The persecutions of the Duke of Alva drove refugees from Flanders to England in 1585, and again a century later, the insane act of the French Government in the Revocation of the Edict of Nantes had a similar effect, and was the means of sending a colony of French Protestants or Huguenots to our shores. Some seventy thousand of these latter entered our country, many of whom, like their Flemish co-religionists were silk-weavers, and these all carried on in the land of their adoption the art in which they were proficient, thus giving a great impetus to the silk manufacture, and rendering a valuable return for the liberality which had welcomed them as refugees from despotism and oppression. One of the most numerous colonies of these exiles settled in the district called Spitalfields, in East London, where many of their descendants still reside.

Some seventy years before the arrival of the Huguenots, however, James I. had been anxious that silkworm rearing should be carried on in this country, and for that purpose caused thousands of mulberry trees to be brought over from France, and planted throughout the country. Great efforts were made to ensure success, and we read that in the year 1629 Walter Lord Aston was appointed " to the custody of the garden, mulberry trees and silkworms near St. James's in the county of Middlesex." But, notwithstanding the royal patronage, or possibly because of it, the attempt shared the fate of so many of James's schemes, and turned out a failure; within a few years the mulberry garden became, to use the words of Evelyn the diarist, " the only place of refreshment about the town for persons of the best quality to be exceedingly cheated at," and later on, it disappeared altogether, crushed out of existence by the all-powerful and ever-extending accompaniment of modern civilisation, "bricks and mortar."

to China, conceived the idea, regardless of the interests of their native country, of watching carefully all the processes of silk culture, and then endeavouring to obtain a supply of silkworms' eggs, and, if possible, transporting them to Europe. Stimulated by the promise of great rewards on the part of the emperor—a somewhat necessary provision, for they knew they were running great risks to life and liberty—and actuated, so it is said, also by a feeling of resentment that so valuable an industry should remain the exclusive property of " unbelievers," they kept their eyes open, of course saying nothing to any one, and, after a time, thoroughly mastered all the details of silk culture. As winter approached, they managed to obtain a considerable supply of eggs, and well knowing that if they were caught conveying them out of the country they would have to pay the penalty of their lives, they carefully concealed them in the hollow canes which they carried as pilgrims' staves. In this way, they conveyed their precious property throughout the long and tedious journey to Constantinople, where at length they safely arrived.

They at once communicated the success of their attempt to the emperor, who was, of course, delighted, and undertook to establish the art of silkworm rearing amongst his people, promising them every encouragement, if they would give the necessary instructions. So the experiment was commenced. The eggs were hatched, and the young grubs fed on the wild mulberries of the district. They flourished, produced silk, and laid the foundations of a new generation for the following season. This was the beginning of silk culture in Europe, and from this insignificant commencement there sprang the whole of the important industry of silkworm rearing, which, continued from year to year and from generation to generation, has grown to such dimensions that it now gives employment to an enormous number of persons in Italy, France, Spain, Turkey, Greece, and South Russia, and produces an annual supply of raw silk weighing no less than three thousand five hundred tons—representing at 15s. a pound, a value in hard cash of £5,880,000.

Thus by the enterprise of two monks, an important blow was struck at the Persian trade, and a valuable addition made to the industrial occupations of Europe; but the Emperor Justinian, having had so much difficulty in starting the new industry, determined to maintain a firm hold upon it, and therefore, in his turn, confined it to his own district, so that for centuries it was Greece and Constantinople that supplied Europe with all the silk she required. About the beginning of the eighth century, the silkworm was introduced into Spain by the Arabs, but it was not till the twelfth century that, partly through the influence of the Crusades, it spread to other parts of Europe. By Roger II, king of the two Sicilies, some Greek weavers were, in 1130, forcibly carried to Palermo, together with silkworms' eggs and mulberry trees, and there they were compelled to carry on the manufacture for the benefit of Sicily. Hence it gradually spread northward throughout Italy, which at the present day is the chief silk-producing country in Europe, and indeed, is second only to China throughout the world. Florence early became a centre of the silk trade, and in order to ensure success, each peasant in Tuscany was compelled to plant at least five mulberry trees on the land he cultivated. Venice also took a prominent position in the new industry, and at the commencement of the fourteenth century, three thousand people were there employed in it.

Towards the close of the sixteenth century, silkworms were introduced into France, by the exertions of the king, Henry IV. Mulberry trees had for a long time been grown in France, the first ones having been introduced from Italy nearly two hundred and fifty years before, and the king argued that as his country produced mulberry trees, it might just as well be provided with silkworms to feed upon them. He was much assisted in his scheme by a celebrated agriculturist of the time, Olivier de Serres, who was the first Frenchman to publish instructions as to the growing of mulberry trees and the rearing of silkworms. At the suggestion of this man, the king imported from Italy twenty thousand young mulberry trees, and large quantities of silkworms' eggs, and these he caused to be distributed in those parts of France that seemed most suitable for the undertaking. The experiment succeeded, and the industry was fairly established. In consequence, however, of the many political troubles of that unfortunate country, the manufacture has had many ups and downs, and in addition to this, the ravages of disease amongst the silkworms, a quarter of a century ago, almost ruined it. It has, however, maintained its ground notwithstanding all obstacles, and still flourishes, especially in the south, where mulberry trees are grown, and the insect reared on a very extensive scale. Lyons is the centre of the industry.

In our own country we are now so familiar with silk that it is extremely difficult to realise how modern an article of luxury it is, and that less than three hundred years ago it was so rare a substance, that even royalty was compelled, on an emergency, to borrow a simple article of dress made of that material, for none of the kind could be found in the royal wardrobe. James I. when King of Scotland, wishing to appear with all due pomp before so exalted a personage as the English ambassador, was greatly however, is only one stage in its career, and we have now to sketch its complete life history. Suppose we take an individual specimen ; it commences its independent life, as most animals do, in the form of an egg. This egg is a minute object, about the size of a turnip-seed, nearly round, but flattened. When first laid it is of a bright yellow colour, and at that time no trace of such a thing as a "worm" could possibly, by the most diligent scrutiny, be found inside. It seems simply to consist of a little semifluid matter surrounded by a horny skin.

But there are wonderful powers lying dormant in this tiny speck of matter, and within the limits of that small cell there are destined to go on changes more marvellous, if possible, than any that take place during all the rest of its life. Under the influence of a suitable temperature, there is to be formed out of that little speck of living matter, a tiny grub, or caterpillar, which, on reaching its completion, is to eat its way out of its little prison house, and take its place in the world as an item in the mighty crowd of living creatures that daily fulfil their destiny upon its surface. Could we but peep inside that tiny eggshell, and, gazing down the tube of a powerful microscope, watch the changes as they take place, we should be inclined, adapting language applied by Professor Huxley to the development of a different animal, to confess that "the plastic matter undergoes changes so rapid and yet so steady and purpose-like in their succession· that one can only compare them to those operated by a skilled modeller upon a formless lump of clay. As with an invisible trowel, a portion of the mass is divided and subdivided into smaller and smaller portions, . . . and then it is as if a delicate finger traced out" gradually the contour of the body, "fashioning flank and limb into due proportions in so artistic a way, that, after watching the process hour by hour, one is almost involuntarily possessed by the notion that some more subtle aid to vision than an achromatic would show the hidden artist, with his plan before him, striving with skilful manipulation to perfect his work."

But the outward signs of these wonderful changes are few and insignificant; they consist almost entirely in changes of colour. A few days after having been laid, the egg begins to deepen in colour, becoming first brown, then reddish grey, and lastly slaty grey, or greenish, according to the breed. It also becomes depressed in the centre. In this condition it remains all the winter, the gradually falling temperature retarding its further development.

When, however, the temperature rises again towards the close of spring, a new series of changes in the reverse order sets in, and the colour gradually becomes lighter day by day, as the time for hatching draws near. But if closely examined now, the pale tint will be seen not to be uniform, and two dark marks may be detected, a black spot and a brownish curve round the outside. The egg-shell is semi-transparent, and the body of the little grub can be traced as it lies coiled up inside. The black spot shows where its head rests, and the dark curve is the outline of its body. At this time, a slight clicking sound may sometimes be detected proceeding from the egg, arising from the movements of the grub inside.

To extricate itself from its prison, it brings its little jaws into requisition, giving them their first employment in eating a hole through the walls that have so successfully sheltered its tiny life during the long winter months. The hole is always made at the side of the egg, never on the flat surface. And thus it makes its entry into the outer world, in which it has but one main business to perform—the operation of eating. But as it looks round in search of somewhat to try the strength of its jaws upon, we may as well inspect it, and see what manner of creature has been fabricated by the mysterious processes of nature out of that unpromising-looking mass that a few months ago filled the now broken egg-shell. It is a dark, almost black, hairy object of which one is at first puzzled to know which end is head and which tail, and as its entire length is but little over one-twelfth of an inch and it is extra thin in proportion, we find it necessary to use a magnifying glass, if we are to get anything like an accurate portrait.

Bringing a little hand-lens to bear on the tiny being, therefore, we soon learn to distinguish the two ends of the body from one another, for at one end there is a hard, smooth, shining, rounded plate, rather too large for the slender body that succeeds it, while at the other, no such structure appears; and we rightly conclude that the former is the head. Looking along the space between these two extremities, we find a cylindrical body covered with hairs, which closer inspection shows to proceed from a number of tiny pale warts or tubercles, placed in rows. The hairs considerably interfere with a sight of the body itself, which, if they were cleared off, would be found to be marked at intervals with slight constrictions passing right round its circumference, as though there were a series of elastic rings attached inside the skin, and drawing it slightly inwards at those points. Counting up carefully, we should then find that, exclusive of the head, these constrictions divide the body into twelve parts, or segments, succeeding one another in line, though in the three immediately behind the head, the divisions cannot be traced on the *back*, the skin there .being puckered and wrinkled. The diameter of the body does not long remain uniform throughout, but the part immediately behind the head soon becomes somewhat swollen. This is perfectly natural, and is no indication of anything wrong with the part in question. On the underside we find some legs, but we see that they are of two kinds, and situated in two different places. Attached to the segments immediately behind the head are .three pairs of what look like little hooks—one pair, a right and a left hook, on each of three successive segments. These are the true legs. Farther back there are five pairs of totally different organs, situated, one pair on the last segment, and the other four on segments six to nine, the tenth and eleventh carrying none. These are the claspers, or prolegs, as they are sometimes called, the prefix " pro " implying here substitution, and not position— " for " and not "fore"; they do duty for legs, but are only a temporary expedient, and no trace of them is visible when the creature reaches its adult condition. The examination of the structure of these locomotive organs, as well as of the mouth organs, we had better postpone till the caterpillar has grown larger, when it may be much more easily effected ; and indeed, many of the points noticed above will be more conveniently observed at a later stage. In its blackish colour and hairy skin, it differs totally from what its future appearance is to be.

As we shall have occasion, in a subsequent chapter, to refer to the food and the method of supplying it, we need not particularize here, but simply say that the little orphan, for such it is, having lost both its parents months before it was hatched, soon sets to work on the food it finds ready to hand, and rapidly increases in size, filling out its loosely fitting skin, and becoming reddish-brown in colour, till in about five or six days' time, it has reached the limits of expansibility of its first garment, for the skin does not grow with the rest of the body, and something must be done to provide for the increase in bulk that has yet to be accomplished. The grub has now passed through what is technically called its " first age," and before entering upon the second, it has to undergo the somewhat trying process of " moulting," or casting its skin. Succeeding ages are passed through with similar regularity, though the exact duration of each will vary with the temperature, and it will generally be found by amateurs that the times here mentioned will be slightly exceeded.

As the period of the change of skin approaches, the silkworm loses its appetite, and soon ceases to eat altogether. It then takes up a position at some chosen spot, raising the anterior part of

2

the body and holding on firmly with its claspers, and assisting its grasp with some silken threads run backwards and forwards across the object to which it is clinging. In this position it remains (Fig. 1) perfectly immovable, and hence its condition is described by the French as "sleep," the caterpillar being said to "go to sleep." When in this state, it is highly irritable, and resents any touch by petulant jerkings of its fore-part. A slight change in the appearance of this part is perceptible ; it seems to become more swollen and misshapen than usual, and the head appears to be thrust forward beyond its natural position (Fig. 2). After the worm

Fig. 1.—Position of Silkworm before moulting.   Fig. 2.—Head of Silkworm about to moult.

has remained in this position some time, perfectly regardless of all that is going on around it, the skin by the head begins to split, the split extends, and presently, by means of working its head from side to side, the front part of the body is dragged through the opening, the legs being drawn out of their skins like fingers from a glove. By muscular contractions and contortions the old skin is gradually pushed backwards off the whole body, and is left in a collapsed condition in the spot previously occupied by the caterpillar, being retained there by the skin of the prolegs and the silken threads.

When the grub has thus wriggled out of its skin, it shows that it was fully prepared for such a remarkable performance, by appearing clad in a new skin, which had been gradually formed beneath the old one at the expense of some of the food that had been ravenously devoured during the last few days of its first age. This second skin is really larger than the first, though it was contained within that, and thus, as the pressure of the old tight-fitting garment is removed, the body expands, and we have the curious spectacle of a creature, bereft of its skin, larger than when it had it on. It has, however, considerably altered in appearance by passing through this crisis (Fig. 3).

Fig. 3.—Silkworm —Second age.

It is paler, of a greyish colour, and much less hairy, only the merest trace of hairs remaining, and we can now see clearly, at the end of the body opposite the head, a prominent object, curving upwards and backwards, in the form of a central horn.

At first sight it seems to be situated on the last segment, but closer examination shows that it is really on the last but one. This horn reminds us of a similar and more prominent appendage in the caterpillars of the hawk-moths, which are not very distantly related to our insect.

Its second age is now entered upon. The period of unavoidable fast has to be made up for, and the pangs of hunger prompt the newly arrayed grub to seek again its favourite food. This it soon does full justice to, with a result similar to that of its first age,—rapidly increasing size and paling colour. This age does not last quite so long as the first, for, about four or five days after the completion of the first moult, the time for the second comes round. This is accomplished in a similar way to the preceding, and the third age is (Fig. 4) entered upon. Again we notice an increased

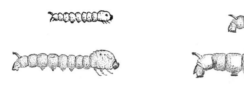

Fig. 4.—Silkworm—Third age.   Fig 5.—Silkworm—Fourth age.

size, and a paler colour, and by this time we can pretty easily detect some curious markings on different parts of the body; on the front of the second segment there is a dark line with generally a small, dark spot on each side of it; on the fifth there are two dark, crescentic marks; and on the eighth, there are two raised, dark spots. The horn at the hinder end of the body is also a more prominent object. The third and fourth ages (Figs. 4 and 5) last from five to six days each, and during this time the caterpillar increases very considerably in size, and becomes much paler.

The period of the fourth moult is an unusually trying one, and not unfrequently individuals perish through their inability to

Fig. 6.—Full-grown Silkworm.

struggle through it. When this moult is over, the animal is very large, and of an extremely pale and sickly-looking colour. Fig. 6 represents the animal towards the close of this age, when it is full grown.

It is now sufficiently large to permit of our easily examining certain niceties of structure that we have hitherto omitted, but it will still be useful to employ a hand-lens to assist in this examination. Commencing with the head, we notice that this is a rounded, polished surface, divided into two parts by a line down the middle. These large polished pieces are sometimes, from their glistening appearance, supposed to be eyes, but that is an entire mistake. There are eyes, it is true, but they are minute, and are placed very near the jaws—an extremely significant arrangement. They will be seen at each side as a number of tiny rounded and glistening black spots. They do not appear to be of any very great use to their owner, who, though not absolutely blind, can yet apparently do little more than distinguish light from darkness. But this is no great drawback, for it will spend the whole of its caterpillar life within an area of a few square inches, and it needs to see only one thing, viz., mulberry leaves, and these its human owner will provide it with.

Between the two polished halves of the head below is a triangular space called the *clypeus*, and below this is a sort of lid stretching across from one side to the other, and hinged above to the clypeus, so as to be movable outwards upon its own upper edge. This is called the *labrum*, or upper lip. It partially conceals the very powerful jaws, or *mandibles*, which, to a creature so devoted to eating, are among the most important organs it possesses. They are a pair of short, but broad and very strong blades, jointed to the head just behind the labrum, working sideways and meeting in the middle, like a pair of shears. The edges which meet are furnished with pointed projections or teeth, so that when the mandibles are closed upon the leafy food, the latter is nipped off in little pieces by these teeth. Here we may notice that these jaws are, in every possible respect, the exact opposite of our own. Ours are placed inside the mouth, where they are sheltered by fleshy lips, those of silkworms are outside, and there are no such things as *fleshy* lips at all. Ours have their hard parts inside, and are covered by the tender, fleshy gums ; theirs have the hard parts outside, and gums are totally absent. The upper jaw of a human being is fixed, and the lower jaw moves vertically towards it, and if he were to eat a leaf, he would place it horziontally between the two. The jaws of silkworms are both of them movable, the movement is lateral, and when *they* eat a leaf, it is placed vertically between the jaws.

The task of getting the food into the mouth is not left to the mandibles alone ; another pair of jaws comes to their assistance— a pair of manipulators rather than biters. They are placed behind the mandibles, and are called *maxillæ*.   They have not the power of the mandibles, being far less strong and hard, and less freely movable, and thus are used chiefly to direct the course of the bitten pieces till they enter the mouth, in which function each maxilla is assisted by a jointed organ called a *palpus*, attached to its side.

Between the two maxillæ is an organ called the *labium*, or lower lip, in the middle of which, below, is a rounded projection, the *spinneret*, from a minute hole in the centre of which the silk issues, when required.   The mouth itself is a small opening situated between the mandibles and maxillæ, and is bounded below by the labium, and above by the labrum.   When food is taken, the leaf is held between the front legs, with its edge turned towards the body ; then the head is raised till it reaches as far up the edge of the leaf as possible ; the mandibles are then opened and brought down upon the leaf with a powerful stroke, then opened again while the head is caused to descend a little, where-upon another snip is taken, then another and another in quick succession, the head being moved a little down towards the under surface of the body after each bite.   Meanwhile, the maxillæ and their palpi do their work behind, and the snipped-off pieces are rapidly passed into the mouth.   From this it is evident that the bitten edge of the leaf will present the form of a semi-circle, or some smaller, but regular, curve.   When one series of bites has been finished, the head is again raised, and another series of nibbles made in curved fashion as before, and so on till the meal is finished.

One more pair of organs is all that remains to complete the the equipment of the head.   These are the two *antennæ*, which are small cylindrical bodies, terminated by a long hair and jointed to the head near the eyes.   They appear to be organs of sense of some kind.

Down each side of the body, there will be noticed a row of nine small oval black rings, one on each segment, except the second, third, and twelfth.   These enclose little openings into the body, which constitute the entrances to the breathing organs.   The black ring surrounds a white space, down the longest diameter of which runs vertically a straight slit ; this is the entrance referred to.   Caterpillars, curiously enough, take in the air they require for breathing purposes, not through any part of the face, as we do, but through these slits in the sides, which are therefore called *spiracles*, *i.e.*, breathing holes, and the dark spots that surround them are called *stigmata*, each being a *stigma*.   These minute slits are the entrances to little tubes, or *tracheæ*, which run hither and thither over the body, branching as they go, and serve as channels by which air can obtain access to all parts of the organism.   The mere muscular movements of the body are sufficient to effect the necessary change of air in them.

Of legs, as we have already said, there are eight pairs, three pairs just behind the head, and five others of a totally different

a   b

Fig. 7.—Legs of Caterpillar, *a*, a single leg, *b*, a pair.

form farther down, the last pair being situated on the last segment. The first three pairs, which are the only true legs, are little five-jointed things (Fig. 7), the last joint being sharp-pointed and curved inwards.   They are very useful for holding and steadying

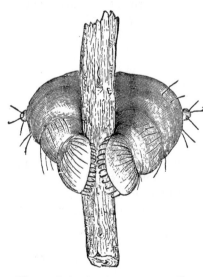

the food while the jaws play upon it.   The other five pairs, or prolegs, are not jointed, but are fleshy, cylindrical pillars, with a circle of strong, curved hooks set around their free end (Fig. 8).   These require the aid of a lens to make out the form and arrangement; and when they are thus viewed, there will cease to be any wonder at the extraordinary tenacity of grip which the caterpillar can mani-fest in clinging to any support. The last pair are rather differ-ently shaped, because they are scarcely separate from the seg-ment on which they are placed, and of which they seem to be a mere continuation.

Fig. 8.—Pair of prolegs of a Caterpillar clasping a twig.

During its fifth and last age the caterpillar eats far more voraciously than ever, and grows rapidly and enormously.   After upwards of a week's incessant feeding, during which it has eaten more than in all the rest of its life put together, and has acquired a length of three inches or upwards, it begins to change colour, turning yellowish, and putting on a waxy appearance, the skin becoming almost transparent. It nibbles its food and scatters the fragments about without attempting to eat them ; it appears restless and begins to wander from the well-supplied pastures it has never before shown any inclination to leave. It moves its head about in all directions apparently feeling for something, which its instinct suggests to it, ought to be near at hand. It begins to climb every obstacle it encounters, being especially anxious to work its way " upward." It disdains its old litter, and " Excelsior " is its watchword. At the same time it is continually producing silk from its spinneret and attaching threads of it to all objects around. All these signs indicate that the spinning time is at hand. The caterpillar is now desirous to provide a retreat for itself in which to perform the last two changes of its skin— moults, each of which will produce a far more startling and remarkable alteration in its appearance than either of the preceding has done. The first is the change to a chrysalis, the second, the change to a moth. It has now to look forward to from three to five days of incessant toil, while it abstracts from its own body the material necessary to enshroud itself.

It is not long in selecting a spot suitable for those momentous epochs in its life, and at once begins to lay the foundations of that beautiful cocoon, which is so highly valued by human kind. Its first operation is to run threads of silk backwards and forwards in an irregular way, from one support to another, at the extreme limits of the space it has decided to occupy, and to continue them farther and farther inwards till only a small oval space is left in the centre, not more than an inch and a quarter long, in which it can only just turn round. Thus far it has not commenced the cocoon proper, these threads being only the outworks, as it were, of its snug fortress, intended to support it and fix it securely, and the actual form and arrangement of the mass will depend entirely upon the exigencies of the position selected. When this " floss," or " refuse silk," as it is called, is finished which will be the case after some five or six hours, the caterpillar devotes its atten- tion exclusively to the oval space it has thus enclosed, and begins with much more care to run a continuous thread backwards and forwards upon the walls of this till a very compact, tough and perfectly opaque, hollow, oval structure is formed (Fig. 9), in the centre of which the caterpillar remains bent almost double, or twisted in the form of an S, so cramped is it for room. This is the true cocoon or, as it is commercially termed, the "pod," the part which is used for obtaining the best silk, and all that man has to do is simply to unwind that long thread which the caterpillar is here twisting and winding about in all directions as fast as it is produced.

The silk is formed inside the body of the silkworm, as a kind of gum which hardens on exposure to the air. When the little labourer wishes to commence spinning, a drop of this gummy substance oozes from the perforation in the spinneret, which is then pressed upon the spot from which the thread is to start. It instantly adheres, and then the caterpillar, moving its head away, draws the gummy secretion out in the form of an exceedingly fine thread, till the next point of attachment is reached, and the thread caused to adhere as before, and so the process goes on till the requisite amount has been produced. In forming the compact portion of the cocoon, the silkworm does not wind the thread

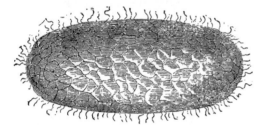

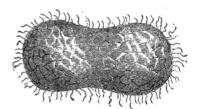

Fig. 9.—"Pod" of Silkworm cocoon.   Fig. 10.—Strangulated cocoon.

round and round, as might be supposed, but, placing itself with its legs outwards, and then bending its head over its back till it is almost double, it sways its head backwards and forwards, continuously tracing a series of figures of eight, the coils being laid across one another, and adhering together by reason of the natural stickiness of the thread. When it has laid a considerable length in one spot, the indefatigable worker shifts its position, and performs the same operation in another, and so on, without ceasing its toil, or breaking its thread, till the whole cocoon is finished. From this it will be seen that the cocoon is composed of a series of patches, as it were, all stuck firmly together, and consisting of one continuous thread, but each patch, nevertheless, formed of an overlying series of loops in the shape of a double curve.

The length of thread thus produced depends upon the size and vigour of the insect, and upon the amount and nature of the food it has eaten. In the best breeds when most carefully reared an enormous length is formed, ranging from seven hundred to one thousand three hundred yards; in other words, the thread of a single cocoon is frequently upwards of half a mile in length. The fineness of the thread will easily be understood when we learn that the cocoon containing this great length weighs on an average, including the enclosed chrysalis, only twenty-five grains.

The caterpillar takes about three days to complete its task, and very exhausted, one would imagine, it must be when it has finished. A calculation has been made that it moves its head backwards and forwards about 300,000 times during the formation of the cocoon, and yet, of course, it takes no food all the while, to recruit its energies or refresh itself after its severe labour.

The cocoon is usually oval in shape, with rounded ends and parallel sides, but frequently it is more or less constricted in the middle, or, as the French say, "strangulated" (Fig. 10). The colour is white in the best breeds, but it is often more or less of a yellowish tinge, and not unfrequently of a deep golden yellow. Different breeds form cocoons of different colours, some being greenish and others even roseate. It is a curious fact that the colour of the prolegs of the caterpillar during its last age always corresponds to that of the cocoon it will spin; so that, by observing these, we may tell what coloured cocoons we shall get.

When the cocoon is finished, the caterpillar may rest in peace. Its active labours are over, and it has now but to resign itself to its destiny, and await its transformation into a chrysalis. It shrinks very considerably in length, but expands in breadth in the middle of the body; its prolegs shrivel up, its true legs are curled inwards, and it becomes stiff and helpless, lying against the side of the silken bed it has just made for itself. Thus it remains for three or four days more, while its skin is gradually parting company with the portions of its body that lie immediately beneath it, and the covering of the pupa, or chrysalis, is being formed between the two.

When this is ready for use, the old skin splits near the head as on former occasions, but this time discloses, not another caterpillar similar to the present one, but a pale object rounded in front and slightly tapering behind, which, though totally devoid of limbs, manages gradually to push its enveloping shroud backwards, till it has quite worked it off behind. The pale object thus revealed is the pupa, or chrysalis. The cast skin must, of course, now lie by its owner during all the next stage of its life, as there is no means of getting rid of it: but as it shrinks to very small dimensions, it does not get in the way, and remains at the hinder end of the chrysalis.

The chrysalis itself is a curious mummy-like thing. It is both somewhat stouter and far shorter than the caterpillar from which it sprang, so that it finds in the cocoon which was a tight fit for the worm that formed it, more than enough room for the accommodation of its altered bulk. It has no legs and no mouth, so that it is unable either to walk or to eat. Its only power of movement is to twist about the hinder parts of its body slightly. Immediately after its exclusion from the caterpillar skin it is pale yellowish white and soft, but it soon hardens and becomes darker, till at length it assumes a reddish brown colour. Two divisions can be easily recognised in its body. There is first the anterior half, on which one can trace, more or less distinctly, the outlines of the fore part of the moth which is ultimately to issue from it, the forms of its eyes, antennæ, wings, and legs being traceable; these latter are all bent down side by side towards the under surface of the body. These are, however, as yet but prophecies, merely the outward indications of what is ultimately to be found within; there are no such things as antennæ, legs, and wings now, and these markings on the pupa skin are, as it were, only the outside of the moulds in which they are to be cast. The posterior half consists of a series of distinct rings, which lessen in size as they approach the end of the body; these correspond to the abdomen, or body, of the adult insect.

The pupa is perfectly helpless, it has no power of self-defence, and, indeed scarcely seems alive; and, if it were in the wild condition would, but for the cocoon in which it has taken the precaution to enwrap itself, easily fall a prey to any one of the myriad enemies that would be continually seeking its life. The outward quiescence is, however, no criterion of the state of affairs within; there all is activity and rapid change; out of the soft and almost fluid contents of this mummy there is to be fashioned, in a little more than a fortnight, a full-fledged moth, a winged creature, not only totally unlike either caterpillar or chrysalis in appearance, but possessing a very important power which they did not, viz., that of reproducing its kind. This is the explanation of that otherwise inexplicable phenomenon, the insatiable appetite which is characteristic of the final age of the caterpillar; it was then taking in supplies destined to become a store of force, to be drawn upon during this period of seclusion and outward inactivity, for the formation of the winged form which represents the final stage in the insect's life.

When the time arrives for the emergence of the moth from its pupa case, the latter, through pressure from within, splits down the back and along the sides of the wing covers, and the anterior part thus divides into several pieces, and allows the inclosed insect pretty easily to wriggle out. On getting clear, it is seen to be in the form of a six-legged four-winged creature, whose wings, however, are soft and wet, and extremely small. But though the insect has escaped the confines of the hard brown case that so closely invested it, it is still a prisoner; it is cut off from the outer world by the compact walls of the cocoon, through which there is no passage, and which certainly will not, in their present condition, yield to any amount of pressure that the poor limp thing within can exert upon them. Moreover, the moth has no jaws with which to bite its way through; how, then, is it to escape? The problem seems difficult, and yet it is solved in a very simple way, though naturalists were for a long time puzzled as to the means employed. From two glands in the head the moth pours out upon the end of the cocoon against which its fore part is placed, a small quantity of an alkaline liquid which possesses a solvent power, not upon the silk itself, but upon that gummy substance which invested the threads at their extrusion and caused them to adhere to one another. In this way the threads at this point are in time loosened, and can then be pushed aside from within by the moth, which thus works a hole sufficiently large to permit of its escape, without intentionally breaking any of the threads, though possibly some few of them may get broken during the struggle. In squeezing its way out, it is greatly assisted, if it can find any rough objects near at hand to which it may cling with its fore-legs, and thus obtain leverage to aid it in dragging its body out.

It now seeks the nearest possible projection from which to hang with its minute limp wings depending vertically behind its back. This is to facilitate the passage of fluid from the body into the minute tubes which form, as it were, the framework of the wings; the latter are thereby distended and speedily assume their proper dimensions, not long after which they dry and harden. Shortly after its emergence, the moth discharges a quantity of a thick reddish fluid, which irretrievably stains everything it touches, and on drying leaves a considerable deposit of a pale reddish powder compacted together. The old pupa case is of course left empty within the cocoon, which therefore now contains two cast-off vestments—the shrivelled larva skin, and the broken pupa case.

The moth is a pale cream-coloured creature with a tolerably stout body, covered with closely-lying creamy hairs, and with four not very large creamy wings, the upper pair rather hooked at the tip, and crossed by faint brownish markings in the shape of wavy lines and bands (Figs. 11 and 12). When fully spread out, they measure a little over an inch from tip to tip; but they are never naturally found in this position; the insect always carries them

sloping away from the body in such a way that the fore ·wing
almost entirely conceals the hind one on each side, and therefore,
at first sight it might be thought that there were but two wings.
In rest the wings lie horizontally, and do not conceal the body.

With care we can distinguish three parts to the body : first, the
head, which is by far the smallest of the three, a small white
rounded part which carries two great black eye-masses, and two
feathery organs, the antennæ, projecting from its sides like a huge
pair of moustaches ; secondly, the *thorax*, a broad and squarish
part immediately behind the head, and having both legs and wings
attached to it ; and finally, the *abdomen* which fills up the space
between the hind wings and projects considerably beyond them,
but carries no appendages. If we compare the moth with the
caterpillar that produced it, and endeavour to trace the corre-

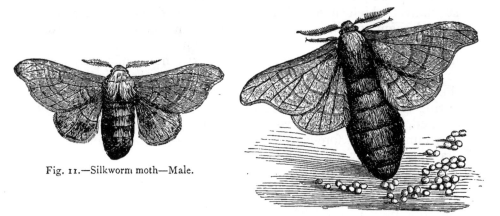

Fig. 11.—Silkworm moth—Male.

Fig. 12.—Silkworm moth—Female.

spondence of parts in these two creatures so utterly unlike, we find
that the head of the moth corresponds to the same part in the
caterpillar, the thorax to the three segments behind the head
which carried the true legs, and the abdomen to the rest of the
body. The segmentation of the body is a good deal concealed
by the long hairy scales, with which the whole creature is covered ;
still with care, we can make out nine rings in the abdomen,
especially when the moth twists that part about. This, it will be
seen, corresponds with the number of segments in that part of the
larva's body which answers to it. By a little hard brushing with
a camel's hair brush the scales may be removed from the body,
and its real structure thus laid bare. It is covered with a brown,
horny skin, which readily shows its composite nature.

The divisions of the thorax cannot very easily be seen on the back, but below we observe the three pairs of legs, each of which indicates a separate segment. The *prothorax*, or first thoracic segment, which is a small one, carries only a pair of legs; the second, or *mesothorax*, much the largest of the three, is better provided, and carries a pair of legs below and a pair of wings above; and the third, or *metathorax*, is similarly furnished, and carries the third pair of legs below, and the second pair of wings above. This disposition of parts is not confined to the silkworm, but is common to all insects; legs and wings are not attached indiscriminately to *any* part of the body, but are always arranged as above on the three thoracic segments.

Few parts of the body are more elegant than the antennæ; they are well worthy of a minute and careful investigation, and indeed, the more minutely we examine them the more do we see to admire. Each consists of a white central axis, carrying along almost its entire length two rows of long blackish threadlike projections, which slope downwards away from the white axis like the sides of the roof of a house from its ridge, the individual threads becoming shorter as they reach the end of the antenna, and dwindling away at last to nothing, so that the organ terminates in a point. This is all that appears to the naked eye. But place the object under the compound microscope, and you will be astounded at the complexity of structure it manifests. The white axis is now seen to owe its colour to tiny snow-white scales packed close along its upper surface, and concealing its really dark exterior. It is not a simple rod, but is composed of a series of fifty minute joints, which, by their power of moving upon one another, give extraordinary flexibility to the whole. Each of these carries two long thin projections, far longer than itself, which are themselves also doubly fringed with hairs of extreme tenuity. These long processes constitute the double row of threads spoken of above. Antennæ of this kind are said to be "pectinated," *i.e.*, comblike. Conjecture is vain as to the use of these organs. They are no doubt organs of sense of some kind, but of what sense or senses is a great puzzle. Several different functions have been assigned to them, with varying degrees of probability, and it is impossible at present to come to any definite conclusion in the matter.

The antennæ are attached to the head just outside two great black hemispheres, which are evidently the eyes. The surface of these, under magnification, reveals a network of hexagonal divisions, each of which is the outer boundary of a complete organ of vision, so that these two rounded masses are not, what at first sight they seem to be, simply two enormous eyes, but each is composed of many hundreds, and hence the name " compound eyes," by which they are known. If we apply to them notions derived from the knowledge of our own eyes, and imagine that they are anything like what we are ourselves accustomed to, we shall be grievously in error. If they were constructed at all like ours we should be able, to some extent, to argue from our own experience, as to what is the nature of insect vision. But, unfortunately, this is not the case, and it is very difficult for us, who have never had any experience in the management of compound eyes, to understand what sort of vision they can produce. We are accustomed to think of an eye as a delicate, somewhat yielding, globular object, snugly quartered in a hollow fitted up for its reception with paddings of flesh, and capable by means of elastic bands of flesh attached to its sides, of pretty free movement, noiselessly and without friction, in the well lubricated cavity. Nothing, however, could be farther than this from presenting a true picture of an insect's eyes. They are not sunk in sockets as ours are, nor are they movable; there are no eyelids, nor any other kind of covering, by which they can be withdrawn from the action of the light, so that their possessor cannot perform such a very simple and familiar operation as " shutting its eyes." Neither is their surface soft and moist, as with our own; they are hard and dry, and fully exposed to the glare of the light, and unquestionably, their surface is not acutely sensitive to the touch, so that no such exquisite torture as we are ourselves sometimes subjected to, can be produced by the contact of foreign bodies.

And yet these eyes are exceedingly complicated structures. Each of the hexagonal facets, as already mentioned, is the outer surface of a single organ of vision. It consists of a piece of transparent, horny substance, beneath which, and pointing towards the centre of the spherical eye-mass, is a curious, elongated body called a crystalline cone and spindle, which is itself surrounded by a dark-coloured material. Thus we have hundreds of these structures lying side by side, all placed perpendicularly to the surface, and all tending to converge towards the centre of the spherical mass, where is placed a large expansion of the optic nerve, one of which proceeds on each side from the insect's brain.

An insect's compound eye is a very beautiful object, when examined under the microscope with a low power. If it be viewed as an opaque object, with a bright light directed upon it by a " bull's-eye," all the facets show up wonderfully distinctly, and present a most attractive appearance. But the clear outer layer may be separated from the underlying coloured parts, and then, when properly mounted, looks something like a piece of network, with the six-sided meshes all accurately formed, and covering the whole area.

Such, in very brief outline, is the nature of the eye of the silkworm moth.   We see that the convexity of form and the prominent position of these eyes are advantages which counterbalance the disadvantage of their immobility.   Though the insect cannot move its eyes, yet it can receive light from many directions at once, in consequence of the peculiar form and position of those organs, and it is, no doubt, able to see far better now than when it was a caterpillar.

The area between the eyes is densely covered with beautiful cream-coloured hairs, brushed down, so to speak, over the forehead.

The head of a moth usually carries various mouth organs, but that is not the case with our silkworm moth.   This poor creature has no apparatus for taking food; it could not eat or drink if it were to try, and, as a consequence, it endures a perpetual fast, not only while a pupa, but in the perfect state as well; its last meal, like its first, is always taken when it is a caterpillar, and though it lives for nearly a month after beginning to spin its cocoon it takes no more food, but carries on its vital processes at the expense of the stores of nutriment laid up in its body during its larval life.

The thorax may be regarded as the central mass of the whole organism, for to it are attached the head in front, the abdomen behind, the wings at the sides, the legs beneath.   The prothorax, when denuded of hairs, appears on the upper surface as a narrow plate.   The long, creamy hairs attached to it stand almost erect, forming round the neck above a sort of Elizabethan frill; by a parting in the middle they are divided into two sets, sloping in opposite directions.   They can be easily distinguished from the rest of the thorax, and indeed are often mistaken for a part of the head.   Below, this segment is much more largely developed, as it there carries the fore-legs.   The mesothorax, which occupies the greater part of the upper surface of the thorax, carries at its sides two creamy tufts or lappets of hairs, which are pretty easily distinguishable from those that cover the central part, and serve to conceal the points of attachment of the wings.

The legs are all six pretty much alike.   Each consists of five parts, only four of which, however, can be easily identified.   First there is a short cylindrical piece called the *coxa* or hip; this is jointed to a hole in the horny covering of the thorax beneath, in which it works.; it is succeeded by a minute joint, the *trochanter*, which cannot be seen till the leg is bared of scales, when it appears as a triangular piece attached obliquely to the end of the coxa; next comes a much larger and rather stout cylindrical piece set on at an angle to the coxa, and loosely furnished with long hairs; this is the *femur*, or thigh; this is succeeded by a curved thinner piece, the *tibia*, or shank, carrying a dense brush of hairs at the side and two movable spurs at the further end; lastly comes the *tarsus*, or foot, which is itself composed of five joints, of which the first is by far the longest. The last joint of the tarsus carries a pair of curved claws, by which the moth can cling to any support that exhibits the slightest roughness of surface. All the tarsal joints are covered above with creamy scales, which give them a curious appearance, reminding one of the feathers down the legs and toes of Cochin China fowls. Between the two claws is a minute pad. When the moth walks it supports itself on its flexible tarsus, clinging with its claws to any irregularity of the surface, and placing its shank and thigh at an angle to one another.

The wings are really composed of a delicate transparent skin or membrane, strengthened and kept in position by a number of fine tubes, called *nervures*, branching out from the point of attachment to the thorax; it is in these tubes that the fluid referred to above at first courses for the extension of the wings, but in the fully formed moth they are dry and stiff. The nervures do not, as might be supposed, branch about the wings in an indiscriminate way; they always follow a definite course, and are constant in number, so that it has been found possible to assign names to them individually, and to the areas into which by their branchings they divide the wings. The arrangement of the nervures, or the neuration, as it is called, varies a good deal in different groups of moths, and is therefore largely employed as an aid to classification. The name "nervure" is rather an unfortunate one for these minute tubes, suggesting as it does some affinity with nerves, with which, however, they have nothing whatever to do.

The clear, membranous wing is completely covered, above and below, with a vast number of exceedingly minute scales, of very varied shapes, but so small, that to the naked eye they appear simply as so much fine dust. If the wings be roughly handled, they readily come off on the fingers. It is these scales that produce the pattern on the wings, which, without them, would have no more adornment than a clear piece of window glass. Each scale is a flat body, terminating at one end in a single point, and at the other in a number of pointed projections, of different lengths, according to the part of the wing from which the scale is taken. The single pointed end is that by which the scale is attached to the wing. In the membrane there are a series of tiny pits, which can be easily seen with a microscope when the scales are removed; into these the pointed ends of the scales are inserted, and there is no other place of attachment than this. The scales are laid so close together that they lap over one another like the slates on the roof of a house. The wing may be denuded of its scaly covering, and its transparent nature thus clearly shown by careful brushing with a camel's hair brush, but there is considerable danger of tearing the tissue of the wing in the process. By chemical operations, the wings may be bleached, and rendered transparent, and then the arrangement of the nervures becomes manifest. To do this, the wings are soaked for a time in strong alcohol, so as to absorb all grease. Then they are laid in a solution of bleaching-powder, or chloride of lime till the colour disappears; they must not be left too long in this solution, or they will become damaged. Next they are transferred to a weak solution of hydrochloric acid for a short time, and then from that to pure water. After being well washed, they may be dried, when they will be seen to be transparent, with the bleached scales still in position. At the outer edge of both wings the covering of scales terminates in a kind of fringe, made of more elongated scales. The hairs which cover the body pass by insensible gradations into the scales which cover the wings.

At the upper corner of the hind wing, close to the body, there is a thorn-like spine, which is a characteristic feature of moths, but is not found in butterflies. In those species that fly, it serves to hook together the fore and hind wings, which thus present a more unbroken surface to the air in the down stroke of the wing. In the present insect, however, it has now no such use, as the silkworm moth has altogether lost the power of flight, and occupies itself merely with aimless flutterings, in which the wings are kept separate.

Such is the creature which, by means of a series of most romantic changes, the wonderful processes of nature have elaborated out of the tiny slate-coloured egg with which the cycle commenced. By far the greater part of its existence has already been passed through, and it has now not much more than a week to live. The great business of perpetuating its race is all that now lies before it; when this is accomplished, its mission will have been fulfilled, and it will perish. As soon, therefore, as the moths have dried their wings after exclusion from the chrysalis, the males seek their partners, and the pairing takes place without delay.

The sexes differ chiefly in the narrower body and more deeply pectinated antennæ of the male. They remain coupled often for many hours. The female then deposits her eggs one by one; these, as they are extruded, are covered with a glutinous secretion which causes them to adhere to the substance on which they are placed. In some breeds, however, they are extruded dry, and therefore do not adhere, but remain loose. In the former case they are laid side by side, and not piled in heaps. Each female, if in vigorous condition, may be expected to produce about 300 or 400 eggs, generally in batches, and the whole period of laying lasts about three days. At first the eggs are bright yellow, but in a few days, if they are fertile, they change to the familiar lavender or slate-colour. Of course, if they are not fertile, they remain yellow, and this is one way of distinguishing those eggs that will produce caterpillars from those that will not. The female will deposit eggs, even if she has not mated, but of course such eggs are rarely of any use. Very occasionally eggs laid by virgin females will mature, and produce caterpillars, the process being called "parthenogenesis," but the proportion of such instances is exceedingly small. For example, it is recorded that M. Jourdain found that, out of about 58,000 eggs laid by unimpregnated females, many passed their early stages of development within the egg, thus showing that they were capable of self-development; but only twenty-nine out of the whole number produced caterpillars.

CHAPTER III.

THE SILKWORM—ITS INTERNAL STRUCTURE.

IN the preceding chapter we have endeavoured to gain a clear idea of the external appearance of the silkworm in all its four stages of egg, caterpillar or larva, chrysalis or pupa, and perfect insect or imago. But if we stopped here, we should have but a very imperfect knowledge of the creature's organization, for all the secret processes of its life are performed inside, and if we would really understand something of the way in which it lives and moves, and performs its various functions, we must pursue our investigations into its interior—in other words, we must dissect it. Packed away within the limits of its body walls there are a great many different kinds of apparatus, and as these lie one upon the other, they need to be separated, and to be set out distinctly each by itself, if their form is to be properly seen and their nature fully comprehended. It is this separation of the different sets of organs from one another which is the object of dissection. The process is a very simple one, and, with a little care and delicacy of manipulation, is readily carried out. If properly conducted, it is not, as some might suppose, in the least degree a repulsive process, but, on the contrary, structures of extraordinary delicacy and beauty are revealed, and a carefully performed dissection will be found to possess considerable artistic elegance.

The study of the internal organization of the silkworm will occupy us in the present chapter, but before we enter upon it, it will be well to endeavour to form some idea as to what we may expect to find in the interior of an animal standing, as our silkworm does, fairly high in the scale of organization. In order to repair the waste that is ever going on in its body through the performance of its vital functions, and to supply materials for its destined increase in bulk, an animal needs to take in food, which is the raw material out of which its frame is constructed. But the food as taken in, though consisting chiefly of the right materials, yet has them not in proper forms of combination, and its chemical composition therefore needs to be altered ; the animal will consequently require a set of apparatus, more or less complex, first for the reception of the food, and secondly for its transformation into a fluid of suitable chemical composition to enable it to act as a body-forming and waste-repairing substance. This set of apparatus is called the *digestive* system. But this fluid, thus formed, will need distribution to the body at large, else it will be of little use ; for waste and decay go on in all parts, and hence repair must be equally widespread. The apparatus which provides for the distribution of this formative fluid is called the *circulatory* system. In its passage through the body, this fluid not only parts with much of the nutriment it contains, but also receives into itself certain waste products of a gaseous nature whereby it is rendered unfit to perform its task ; these can be got rid of by simple exposure to the air ; there will be necessary, therefore, some apparatus by which the fluids of the body can be brought into contact with the atmospheric air, and this is called the *respiratory* system. Certain other waste products are to be got rid of as liquids, and for this purpose an *excretory* system is required. The animal is, further, to be put into relation with the world around it, by means of some apparatus through which it can receive information from that world, and regulate its own actions accord ingly. The set of organs designed for this use is called the *nervous* system. But this again necessitates another set of organs, by means of which its movements may be effected in response to the requirements indicated by the nervous system, and for this purpose it must have a *muscular* system. Again, there are sometimes certain special products which have to be separated from the fluids of the body, not simply to be got rid of out of the system, but for use in some way or other, either within the body or outside, and the apparatus for this consists of *secreting* glands. Lastly, provision must be made for the propagation of the race, and therefore a set of *reproductive* organs is necessary.

Bearing in mind these principles, let us take the full-grown caterpillar as the most suitable subject for the investigation. The first business is to kill it in such a manner as not to damage it. This may be accomplished either by placing it under a glass with a small piece of cotton wool on which a few drops of chloroform have been dropped, or by immersing it in alcohol, or spirits of wine, for a few moments. A little apparatus will be required for the work of dissection ; but it is of such a simple nature, that most of it can be easily prepared by the operator himself. The dissection must be performed under water, as that medium causes the various organs to float out from one another, instead of lying in one undistinguishable heap, and enables them to be more distinctly seen and more easily separated from one another. A dish of some kind is therefore required. One of the little oval white dishes which confectioners use for potted beef (not the round ones)—say about five or six inches long, or a small rectangular tin box about four inches long, such as chemists sell with glycerine jujubes—will answer very well. This must be supplied at the bottom with a layer of some substance, into which pins may easily be stuck ; solid paraffin is about the best material ; this may be obtained at the operative chemist's, or, instead of this, an ordinary paraffin candle may be used. The material is to be melted (say in an iron spoon over the gas) and poured into the dish till a layer not less than a quarter of an inch in thickness is produced, which will adhere to the bottom ; or the solid substance may be laid in the dish in lumps, and then melted by standing the dish in a pan of boiling water, when it will spread itself out into a suitable layer and attach itself as before. It adheres to the tin box much more readily than to the glazed earthenware dish. It may then be stood aside till quite cold, when it will be sufficiently hard to withstand pressure, but at the same time soft enough to permit of pins being thrust into it.

Cold water is then to be poured into the dish to the depth of about an inch. Now it is ready to receive the body of the cater-

pillar. This must be pinned on to the paraffin, beneath the
water, with a fine pin at each end of the body. Ordinary dress
makers' pins are too stout and not sufficiently fine-pointed to be
conveniently used; the best for the purpose are "entomological
pins" which may be obtained at any naturalist's. In default of
these, short needles, or the pointed ends of broken ones will do.
The caterpillar should be fastened down firmly, and slightly
stretched, or there will be difficulty in keeping it sufficiently steady
to perform the dissection neatly. Now we shall require a pair of
fine-pointed steel forceps, a pair of dissecting scissors with rather
long handles and fine short blades, and a dissecting knife or
scalpel, which is simply a long-handled, thin-bladed, sharp knife.
All of these may be purchased at a cost of a few shillings at any
good optician's. Of course substitutes may be used for the two
latter, such as ladies' fine embroidery scissors, and a small, sharp,
penknife, but they will not perform the work so neatly, nor be so
convenient to use. Most of the following dissections, however,
may be performed by the help of scissors and forceps alone.

In order to make the greatest possible use of a single dissection,
a blank note book, a hard, fine-pointed black pencil, and two or
three coloured pencils should be provided. By means of these,
drawings of the dissection in its various stages may be made, and
will be found exceedingly helpful towards really understanding
the creature's anatomy. If carefully made, they serve for reference
at any time, and are then almost as valuable as the original dissection itself. The coloured pencils are for putting in the different
organs in different colours, thereby to distinguish them from one
another more easily. It may seem at first thought that this
drawing must be a formidable affair, especially for those who have
had no instruction in draughtsmanship, and indeed some little
difficulty will, no doubt, be encountered at first; but it will pass
away by experience, and a little practice will produce a fair degree
of proficiency. In all the drawings great pains should be taken
to make the *outline* of the various parts distinct and correct, and
the relative position of parts as accurate as possible; minute
details need not be attended to at first, the aim being not simply
to make pretty pictures, but to show the true form and relation of
parts, to make the drawing a sort of key to the work, on which,
e.g., the names of the parts may be written. The dissection may
be preserved from one occasion to another, or indeed for an
indefinite time, by simply keeping it in spirit; methylated spirit
will do very well. A pair of dissecting needles, that is, ordinary
sewing needles firmly fastened into wooden handles, such as are
used for camel's hair brushes, may usefully be added to the above

apparatus. They are useful for separating organs from one another in places which the scissors and scalpel cannot reach.

We will now suppose that we have all our apparatus in working order; in front of us, the dish with paraffin layer at bottom and water above it, and the caterpillar firmly pinned down; and at our side, the forceps, scissors, scalpel, and needles, and a plentiful supply of pins. A pocket-lens should also be kept near. If the dissection has to be performed by artificial light, the rays from a lamp should be directed upon the object by means of a bull's-eye, otherwise it will be difficult to see the more delicate parts. Suppose the caterpillar to be pinned out in the natural position of crawling, back upwards. It will be a very good exercise to make our first drawing of the creature as it is before dissection, and to mark on the drawing the names of all the parts that are visible outside. This done, we prepare for the actual dissection. Holding the forceps in the left hand and the scissors in the right, with the former we seize gently a tiny portion of the skin of the back, and with the latter cut a slit in the skin from one end of the body to the other. The slit must be cut in the skin only, and we must be very careful not to injure the parts beneath, so that only the extreme tips of the scissors must be used. Next a transverse slit may be made at each end of this, on each side of it, and the skin then gently separated on each side from the underlying organs by means of the scalpel or needles, the skin itself being still held by the forceps. When it is well separated on each side, it may be pinned down to the paraffin along its edges, the pins being inserted in a sloping manner, and in such a way that their heads shall point away from the dissection, and thus not impede the hands and tools of the operator.

In this way, all the interior of the creature's body will be laid open, but at first little can be made out of the organs it contains. This arises from the fact that a good deal of the space just underneath the skin is occupied by a quantity of a whitish substance arranged in the form of crinkled folds or frills, or something like rather solid lace-work. This is called the "fat body," and according to the degree of its development, it more or less obscures the other parts. It constitutes a store of reserve material, and is found only during the caterpillar stage. It is produced by the extravagant eating of the larva, which takes in more food than is requisite for present needs, and the superabundance goes towards the formation of the "fat body," which will again be utilised during the final transformations of the insect.

In all probability, notwithstanding every precaution taken to guard against cutting through more than the mere skin, the operator will, especially at his first few attempts, have cut through and irretrievably damaged a long, fine tube that lies just under the skin all along the middle of the back, and is therefore called the " dorsal vessel." This is the creature's heart, but as there is some difficulty in recognising it when dissecting from above, we will neglect it for the present, and suppose that we have cut through and spoilt it.

We must now, with the point of the scalpel or with the needles aided by the forceps, gently remove the " fat body " as much as possible from the underlying organs. As we do this we shall find that we lay bare a long tube which runs straight along the body, from one end to the other, and in its diameter is more than half as big as the body itself (Fig. 13). This is the digestive tube, into which the food is received, in which it is digested, and from which the nutriment it contains soaks through to the body at large. It appears of a dark colour ; but this is a deceptive appearance ; it is really a pale tube, but, as it is semi-transparent, the dark mass of the food it contains shows through its walls and produces the dark tint. We shall presently wash out the contained food, and shall then be able to see the real colour of this great and important organ.

Meanwhile, we notice that the tube seems anchored down on each side by a series of extremely fine threads, reminding one of way in which the tiny inhabitants of Liliput pegged Gulliver down to the ground by the hairs of his head, while he was lying asleep on his first arrival in their country. These fine threads, when they reach the digestive tube itself, begin to branch and spread out like fine roots over the sides.

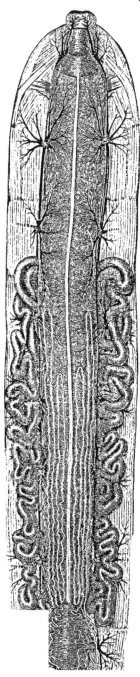

Fig. 13.—Body of Silkworm dissected, showing digestive tube, silk glands, Malpighian tubes, and portion of the tracheal and muscular systems.

These are some of the breathing tubes, and as we shall have
occasion to notice them more particularly presently, we will
now cut them through and so release the digestive tract from
its moorings. If now we cut through the latter just behind
its commencement, we shall be able, by gently waving it about
in the water, to wash out a good deal of its contents, and to turn
it on one side and so examine it more satisfactorily. It is then
seen to be, throughout the upper two-thirds of its course, a very
thin-walled tube, while the hinder third has walls a good deal
thicker than the rest.

The part immediately behind the mouth is called, naming it
it after the corresponding structure in human anatomy, the *œso-
phagus*, or gullet.   The food does not remain in this, but it simply
serves as a passage into the stomach.   Succeeding it comes the
stomach itself, the true digestive cavity, where the coarsely masti-
cated food is received to be submitted to chemical change.   This
occupies by far the greater part of the tube, and indeed fills so
large an amount of the space enclosed by the body walls, that the
caterpillar might, not inaptly, be described as one great stomach.
The enormous extent of this portion of the animal's organization
is, of course, the key to its extraordinary appetite.   The stomach
is succeeded by the *intestine*, the diameter of which is at first less
than that of the preceding part, though just before the termination
of the tube it enlarges again.   In this the undigested remains of
the food are received and stored up, and by the pressure of its
muscular walls compacted into little cylindrical pellets with
grooves down the sides, which are periodically expelled from the
end of the intestine as excrement.   The walls of part of the
intestine are raised inside into a number of prominent muscular
ridges which project into its cavity, and can easily be seen by
slitting open with the scissors this division of the digestive tube.
These ridges are the causes of the grooves down the sides of the
excrementitious pellets above alluded to.   The wide terminal
portion of the intestine is called the *rectum*.

At the junction of the intestine with the stomach there are on
each side six very long and exceedingly fine tubes projecting from
the outside, some parts of which are laid neatly in longitudinal
folds along the hinder part of the walls of the stomach (see Fig. 13),
and others packed closely together like so many tangled threads
just opposite the point of their insertion.   They are the so-called
Malpighian tubes, thus named after a celebrated Italian anatomist.
They are considered to belong to the excretory system, and
indeed, to have a similar function to the kidneys of the higher
animals.   Though they appear simply like threads, they are

really tubes, and they pour the products which are collected within them into the intestine, to be got rid of as part of the excrement.

Lying partly by the side of, and partly underneath the stomach, but quite separate from it, are some more twisted tubes (see Fig. 13).   These are the *silk-glands*, and there is a pair of them, one on each side.   Each consists of three parts; the central, which will be the first seen, is in the form of a stout, yellowish tube, bent in three regular folds at the side of the stomach ; this is prolonged behind into a narrower tube which is much twisted about, and in front into an exceedingly fine tube which runs under the gullet straight towards the mouth.   Before reaching the mouth, however, it is joined by its fellow of the opposite side, and the two unite to form a single canal which terminates in the spinneret before mentioned.   In the dissection there will be no difficulty whatever in following these tubes through the greater part of their course.

The silk is formed as a gummy secretion, in the lower divisions of the tubes, whence it is conveyed in the shape of exceedingly fine threads along the two straight tubes to the canal formed by their junction.   Here the two threads, which of course lie side by side, are bound together into a single one by a gummy secretion which is poured into the canal by two little tubes proceeding from two small glands at its sides.   It is this gummy secretion that gives the silk the beautiful smoothness and gloss which is one of its chief recommendations as a textile material.   Thus it is plain that what appears like a single thread as it issues from the spinneret of the caterpillar, consists in reality of two threads, lying side by side, but united by a sort of varnish into one. Microscopical examination confirms this conclusion, for if the thread of the cocoon be sufficiently magnified, it is plainly seen to be double throughout.   A high degree of magnification, however, is necessary, because the thread, even when double, is of such extreme tenuity that it would require, on an average, no less than two thousand of them laid side by side, to cover a breadth of as much as a single inch.

Underneath the digestive tube lies the greater part of the nervous system.   Though sufficiently large to be able easily to be seen when by itself, it may yet at first be missed, on account of its similarity in colour to the parts over which it lies, and will therefore need careful looking for.   It is a most beautiful structure of great delicacy, and its separation from the surrounding parts will require a steady hand and considerable nicety of manipulation.   To get a rough idea of its form, we may imagine two fine threads of white cotton of equal length, placed side by side, and then, at regular intervals down their whole length, a series of knots tied across the two combined. Thus we should have a kind of chain of knots, or lumps, connected by two parallel threads. This is something of the sort of structure we have to look for. It is to be found all along the middle line of the body (Fig. 14 *b*), immediately beneath the digestive tube, but quite separate from that throughout the greater part of its length. The little swellings of nervous matter, which correspond to the knots on our threads, are called *ganglia,* and therefore the central mass of the nervous system is usually described as a " double chain of nervous ganglia."

By the removal of the digestive tube in the manner indicated above, we shall be pretty sure to have torn and spoilt the beginning of this chain, which is closely connected with that tube. If, however, before removing it, we had examined carefully that part of it immediately behind the mouth, we should have noticed on its upper surface two small, rounded, whitish knobs, very near together, and connected with one another; and a closer examination, aided by the hand-lens, would have shown two delicate, whitish threads proceeding from them, and passing, one from each knob, down the sides of the gullet, until they united with another pair of knobs beneath. Thus a nervous ring is formed round the gullet, with a large pair of ganglia above, and a smaller pair beneath. The larger pair above is sometimes called a " brain," and though this may not be a very suitable name for it, yet it serves to indicate that from this pair proceed nerves to such organs of sense as the creature possesses, and that thus it is to the caterpillar, in some measure, what the real brain of a quadruped is to it. It is this upper pair of ganglia which one can scarcely avoid tearing off when the digestive tube is removed, unless special pains be taken to preserve them. To do this, we must carefully cut through the gullet just behind them, but far enough away to avoid injuring the nervous threads which join the two pairs of ganglia, and then the nervous ring or collar can be slipped off the remaining portion of the gullet without damage to itself, and without being separated from the rest of the chain.

This being done, we notice that though the two threads joining the first two pairs of ganglia are wide apart, to allow of the gullet passing between them, those of the rest of the chain are much closer together, and parallel to one another. Each little swelling on the chain consists of a pair of ganglia, placed close together and united with one another by threads, and from each pair there pass a few fine threads to different parts of the segment in which *a*   *b*

Fig. 14.—Nervous system of silkworm—*a*, Moth, *b*, Caterpillar.

it is situated.   There is a separate pair of ganglia for each segment, so that altogether we can count thirteen of them.   If we
cut the nerves which proceed from their sides, as far away from the
ganglia themselves as possible, we may then raise the whole nervous
chain from its bed, and by slipping a strip of black paper under it,
float it away, and at the same time throw its form into great
distinctness by means of the dark background of the paper.

There will now remain only two sets of apparatus in our caterpillar's body—the breathing tubes or *trachea*, and the muscles.
We will take the former first.   By the time all the preceding
dissection has been accomplished, the tracheæ will have been
considerably damaged, and it will be better to open another
caterpillar if we wish to get a good view of the respiratory system

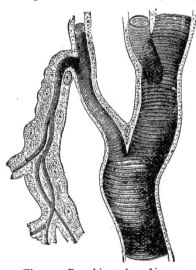

as a whole.   Opening it in the
same way as before, and simply
pinning out the skin without removing any of the organs, we can,
if we turn the silk glands aside,
trace without much difficulty the
course of its main branches.   At
the position of each spiracle there
will be seen a great branching in
all directions of tubes of varying
diameter (see Fig. 13), and running from each of these to its
neighbour above and below a
large, straight, unbranched tube.
Thus, down each side of the body,
in the line of the spiracles, there
is a long tube, with tufts of

Fig. 15.—Breathing tubes of insect.   branches at intervals, along its

course.   Then, as has been pointed out before, numbers of
minute tubes pass in branching tufts up to the walls of the digestive tract and hold it in position by spreading over its surface.
The extreme ends of the tubes will be too fine to be followed
without a microscope ; they penetrate to the remotest corners of
the body, and carry on their beneficial work in all parts of the
organism at once.

The structure of the tracheæ is extremely interesting.   Under
the microscope each appears as a transparent tube, in the interior
of which can be traced a spiral line with its coils very close together (Fig. 15).   This is a stiff elastic thread which serves to keep
the tubes open, and prevents them from collapsing under any
pressure the surrounding parts may cause by the movements of

the body. Of course closure of these tubes would be a serious matter for the insect, the aëration of whose blood depends entirely upon their being kept open and in free communication throughout with the atmosphere ; hence the spiral-thread arrangement, a device which we men have imitated in our non-collapsible india-rubber gas-tubing, which carries a wire coiled up inside. A very pretty sight do these tracheæ present under the microscope, and they have the advantage of being very easily prepared for inspection. All that is necessary is to snip off a little piece of one of the tubes, to place it with a drop of water on a glass slide, and then to cover it with a thin glass coverslip, and it may then be put under the microscope and viewed most successfully.

The *muscles* are extremely numerous, and are easily seen as sets of little, straight, whitish bands, arranged in various directions. Some go straight down the sides, and by their contraction serve to shorten the length of the body—many of these are shown in Fig. 13 ; others pass in various directions obliquely across the segments, and by their united actions serve to alter the diameter or the body. Others move the legs and claspers, and again others the jaws, but these are too small to be dissected out by any but an expert. This well-developed muscular system gives the body of the caterpillar extreme flexibility, and enables it to be bent about in all manner of directions.

One more organ yet remains to complete our account of the anatomy of the silkworm larva ; it is that delicate apparatus which we necessarily destroyed on commencing our dissection, viz., the *dorsal vessel*, or heart. To see this, we pin out another caterpillar, wrong side upwards, *i.e.*, on its back ; now cutting a slit in the skin the whole length of the body, we pin it out, and then remove the nervous and digestive organs, and the silk glands. By this means we lay bare the organ in question. It runs along the middle of the back, just under the skin, where it will be seen as a filmy whitish band. Its structure is rather difficult to make out. It consists of a flattened tube, closed at the hinder end, but passing in front into a narrower tube which is the aorta, or main artery of the body. This cannot, however, be traced very far forward. The sides of the dorsal vessel have some perforations, which serve as entrances for the blood. The whole apparatus is overlaid as we see it in the dissection, *i.e.*, really *under*laid, by a series of pairs of fan-shaped muscles, called the *alary muscles*, whose function is not very certainly known. The blood of the silkworm is not red like ours, but almost colourless. It is not contained in blood vessels, for, with the exception of the rudimentary aorta mentioned above, no such organs are discoverable. It simply fills all parts of the body not occupied by the different organs, but, nevertheless, it does not remain stagnant, but is caused to circulate through the length and breadth of the body. The mechanism which produces this result is the dorsal vessel. If we watch carefully the back of a living silkworm, we shall see down the middle a mark which appears to be just underneath the skin, and is like a bluish or greyish band. It is continually changing its diameter; at one time it contracts till it almost disappears, and then again expands to its full width; but, as the narrowing does not take place throughout its whole length at one time, but progresses gradually from behind forwards, there is afforded the curious appearance of a series of waves coursing towards the head. This is the beating of the heart, and the bluish band is none other than the dorsal vessel itself. From the general cavity of the body the blood enters the heart by the slits in its sides, and then, by the contraction of its walls, it is pushed forward through that organ, and out at the aorta into the body cavity again. On the expansion of the walls of the heart, more blood rushes in from the body, to be again propelled towards the head as before.

The rate of the pulsations depends, of course, upon the health of the insect, which again is affected by the quality of its food. When it is fed on its natural diet of mulberry leaves, the heart has been observed to beat from forty to forty-five times per minute, but when fed on lettuce, which is sometimes used by amateurs as a substitute for the true food, this rate, which indicates the natural and healthy condition of the animal, is reduced to from twenty to twenty-five, showing a considerable falling off in vitality and energy.

The reproductive organs do not exist in any other than a rudimentary condition till the insect becomes a moth. The caterpillar is, functionally, neither male nor female, and has no power of reproducing its kind. On the other hand, the silk glands which are very highly developed in the silkworm, are confined to the caterpillar stage, and in the perfect insect no trace of such a thing can be found. This is, of course, no more than was to be expected. The chief object of the glands is to provide silk for the formation of the cocoon, and almost the only other occasions on which they are brought into play are the moulting periods, just before each of which, as has been already mentioned, they are called upon to provide the few threads required for anchoring the old skin. By the time the cocoon is finished, they are, as a rule, emptied of their contents, and as no more food is taken, of course there is no chance of their being refilled; it is not surprising, therefore, that as there is no further use for them, they disappear altogether in the subsequent stages of the insect. Even in the caterpillar, they are far larger just before spinning the cocoon than at any other time.

In the moth, too, the nervous ganglia become reduced in number (Fig. 14 a) owing to the fusion of some of them together, and those of the head become very much larger. Very considerable changes also occur in the digestive system; if Fig 16, which represents the digestive system of the moth, be compared with Fig. 13, it will be at once manifest that the apparatus has so changed as to be hardly recognisable as the same thing. The gullet has become extremely narrow, and carries a large dilatation, the crop, at its side; the stomach is much reduced, and would scarcely be recognised but for the delicate malpighian tubules at its junction with the intestines; and the latter organ itself has become longer and extremely narrow in front, though greatly dilated behind.

In the foregoing pages we have endeavoured to give a brief outline of the anatomy of our silkworm larva, as it would present itself to one who should proceed to the dissection without any previous knowledge of the subject. To many it will be a matter of surprise that there is so much to be found in the body of a simple caterpillar; but the truth is, that the above is only an exceedingly brief and imperfect description

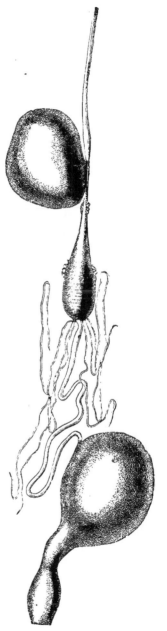

Fig. 16.—Digestive apparatus of Silkworm moth.

of what is, in reality, an organism of considerable complexity. Any one who is curious to know what can really be made of the body of a caterpillar, if the work be thoroughly carried out, even without any great aids in the way of high magnifying power, should endeavour to obtain a sight of a remarkable book which was published in France more than a century ago by a most indefatigable worker named Lyonet. It is a large quarto volume, consisting of a treatise of upwards of six hundred pages, on the caterpillar of the Goat Moth, or, as the author calls it, " Le Chenille qui ronge le bois de Saule," *i.e.*, the caterpillar that eats the wood of the willow. It is illustrated by a large number of plates, in which are depicted with marvellous care and accuracy numerous dissections of the animal, showing the disposition of its digestive, nervous, respiratory, muscular, and other organs with most extraordinary minuteness. If that painstaking observer had lived a century later, and undertaken his work with all the appliances of the modern microscope, he could easily have doubled the size of his book.

We are now provided with the necessary information for considering the position of the silkworm in the animal kingdom. The facts that it possesses no skeleton of bones or other hard parts inside, and that its muscles are attached to the *inner* surface of an *outer* hard covering (the leathery skin of the caterpillar, or the horny covering of the chrysalis and moth), instead of to internal hard parts, show that it belongs to the great Invertebrate section of the animal kingdom, or that division which contains all animals without a backbone or its representative. We have shown that its body is composed of a series of segments, that it has jointed legs, and that its nervous system consists of a double chain of ganglia; these facts indicate that it belongs to the most extensive of all the invertebrate sub-kingdoms, viz., the Arthropoda, where also are located all such animals as crabs and lobsters, shrimps, spiders, scorpions, mites, centipedes, beetles, bees, grasshoppers, etc. Again, it breathes by tracheæ; its body, when mature, consists of three parts—head, thorax, and abdomen; it has at the same time six legs and four wings; these are peculiarities of structure which cut it off from several of the above-named creatures, and indicate that its proper place is in the class Insecta, or Insects, the largest and most highly developed group of all the Arthropoda.

But this class contains an enormous assemblage of creatures of the most varied types, and we must again appeal to the structure of our insect to determine in what order of the class it is to be placed. The two facts that its four wings are covered above and below with feathery scales, and that it passes through the experiences of caterpillar, chrysalis, and moth life, are sufficient to settle this matter, and to show that it is to be referred to the order Lepidoptera (scale-wings), where it finds itself located with all other kinds of butterflies and moths, but with no other sorts of insects. The Lepidoptera, again, are a very extensive order, and are divided into many different groups. To determine to which of these the silkworm belongs, we appeal to the facts that it forms a good silken cocoon, and that its mouth organs disappear when it is fully grown; these peculiarities show that it is a member of the section Bombyces, or silk spinners. From most other of the silk spinners, such, *e.g.*, as the Emperor moths, it differs in that the caterpillar has no warts on its body, but carries instead a horn on the last segment but one, and that the wings are small in proportion to the size of the body. These and some other minute features indicate its family as the Bombycidæ, and in this family, which is one of small extent, it is placed in the genus Bombyx, with a specific name of *mori*, which is the Latin equivalent of the expression " of the mulberry tree," to indicate its attachment to that tree. Thus its full scientific name is *Bombyx mori*, or, as it is now sometimes called, *Sericaria mori*. We·may exhibit its classification in the following table :—

> *Section.*—Invertebrata.
> *Sub-kingdom.*—Arthropoda.
> *Class.*—Insecta.
> *Order.*—Lepidoptera.
> *Group.*—Bombyces.
> *Family.*—Bombycidæ.
> *Genus.*—Bombyx (Sericaria).
> *Species.*—Mori.
> The Common Mulberry Silkworm.

CHAPTER IV.

THE SILKWORM—ITS REARING AND MANAGEMENT.

In consequence of its domestication, the silkworm of commerce, whose structure and life history we have just detailed, has become much altered from its primitive condition, and has developed many peculiarities. The ·most remarkable of these are the absence of all desire to wander on the part of the larva, and the inability of the moth to fly. The great difficulty in keeping

caterpillars in general is that they need to be so closely im-
prisoned to check their erratic habits, and they will frequently
make persistent and untiring efforts to escape from confinement,
and exhibit the greatest ingenuity in finding out the weakest
points in their prison walls and making their exit thereat, to the
great chagrin of their owner when he finds his treasures fled. But
silkworms, fortunately for their possessors, need no such close
guarding, and never manifest the slightest tendency to wander
till the "mounting" season arrives. In its wild state no doubt
the silkworm was as fond of wandering as other caterpillars, but it
has lost all its spirit of enterprise during the long generations of
luxurious life in which it has not had to provide for its own wants,
but has been well fed and well tended by human kind.

Then again, it is a most remarkable thing that the moths,
though possessed of fully formed wings seem either completely
devoid of the power of using them for flight, or totally ignorant
of the method of doing so. They have plenty of muscles to move
them with, but the utmost they can do is to make a rapid
fluttering or fanning motion with them, which is wholly insufficient
to raise themselves from the ground It is incredible that this
should have been the case in their wild condition. It is true
there are many Lepidoptera which in the native state do not fly,
but in such instances, it is usually only the female that behaves
thus, and she has either no wings at all, or only very rudimentary
ones, while our silkworm moth has these organs fully formed in
both sexes. Moreover, *Bombyx mori* belongs to a section of
Lepidoptera, some members of which, the males especially, are
exceedingly powerful and rapid fliers. It is said that if silkworms
be reared in the open air for a few generations, they recover the
power of flight, a further proof that it is the artificial conditions
under which they are now reared that has deprived them of the
muscular energy and vigour which once, no doubt, characterised
them.

In all probability, too, the pale, sickly hue of the caterpillar is
another result of long domestication. The original stock seems
to have had dark brown larvæ, and in every large batch there
appear a few individuals which preserve to some extent this
ancestral characteristic, and appear as dark brown or blackish
brindled. These are what the French call "vers tigrés" or "vers
zébrés." Captain Hutton, some years ago, made many experiments with these dark forms, and endeavoured to obtain from
them a race possessed of the original appearance. He selected
all the dark worms from a batch, and reared them apart from the
rest, allowing the moths to couple amongst themselves only. By

this means, he obtained a batch of eggs which next season yielded caterpillars that were nearly all of the dark brindled form, while the white batch yielded, as before, only an occasional darkie. These latter were then added to the dark stock, which was also weeded of its pale ones, and they were reared separately, as before. Next year the offspring of these were still darker, as well as larger and more vigorous. By proceeding in this way for several generations, he at last managed to get an entire brood of dark worms, which he regarded as the nearest approach to the original appearance of the insect.

From the above, therefore, it would seem that the present race of silkworms is in a pampered and unnatural condition, something like prize bullocks, or fattened pigs, excellently well adapted, indeed, for the purpose their degenerators had in view in thus tampering with their constitution, but altogether unable to take their place and fight the battle of life amongst their wild contemporaries. It is no wonder, then, that the animal is now more subject to disease, and that its rearing requires a good deal of care and attention.

Like most other domestic animals, *Bombyx mori* exists under the form of a great number of varieties, which differ more or less in the colour and size of the cocoons and in the shape of the moth's wings, and no doubt, if suitable selection of parent insects were made, other peculiarities still might be developed. From the colour of their cocoons, the different varieties are known as " yellows," " whites," " greens," etc.

A large number of varieties, differing from one another often only in such minute points.as to be undistinguishable except to the expert, are reared in Europe, and in many instances, each different locality has its own peculiar breed. There are, however, three important and well marked European varieties which can be pretty easily distinguished, viz., the *Milanese*, an Italian race forming fine small yellow cocoons, the *Ardèche*, a French variety with large yellow ones, and the *Brousse*, which is a Turkish stock and yields large white cocoons. The yellow varieties are most commonly reared by amateurs in this country, because they are somewhat more hardy than the white races, but in commerce white silk is the most valuable.

In rearing silkworms, if one wishes to watch their whole life history, of course the first thing to be done is to procure the eggs. If it be desired only to keep a very few, the eggs will no doubt be best obtained from some friend who has already bred the insects and has a stock to spare. But when they are reared for commercial purposes, the eggs must be obtained in enormous numbers, and their selection becomes an important matter, as they and the insects reared from them have a definite monetary value. Throughout the present chapter we shall endeavour to give such information as shall enable the amateur, who takes up the matter solely as a pastime, successfully to rear his charges, but in addition to this we shall describe the method in which the " educations " are carried out in those countries where silk-worm-rearing is a staple industry. Commercially, the eggs go by the name of "seed," and are sold by weight, and in some parts of the world form a very important article of export. Their value is about twenty-five shillings the ounce, and at first sight this may seem rather expensive ; but when we remember that an ounce contains about 40,000 eggs, the product of at least a hundred moths, the price is evidently cheap enough, the rate being about one hundred and thirty eggs a penny.

Some years ago much damage was done to the European races by the rapid spread of some exceedingly destructive epidemics among the insects, and, in consequence, foreign eggs had to be imported, to replenish the stock and improve the breed. Japan was largely drawn upon for this purpose, and still supplies Europe and the United States with great quantities of eggs, no less than six million dollars' worth being said to be annually exported from Yokohama to San Francisco alone. Of course the eggs must perform the long sea voyage during the winter ; they are generally bought in Japan in September and then despatched to their different destinations in the winter months. The two principal varieties thus used are called the White and Green Japanese Annuals.

The eggs are usually laid on pieces of cloth or paper, to which they adhere by their natural gum, but some races have no power of adhesion, and are therefore deposited loose ; such are, for example, the " Adrianople Whites." When kept by amateurs the moths are usually confined in cardboard boxes for laying, and therefore the eggs are attached to the bottom of the box or to paper. For commercial purposes, the eggs are usually detached during the winter ; this is done by washing the cloths gently in water, which dissolves the natural gum by which they are fastened, and they can then be gently removed by the edge of a paper-knife or other similar instrument. The washing does not in the least injure the eggs, provided they are well dried afterwards, that mould may not attack them. If any float when they are in the water, such are considered worthless and are thrown away. The fertile eggs will be of a slaty grey or dark green colour, and any which permanently retain their primary yellow tint are thereby known to be useless. It is possible, however, to impart even to dead or unfertile eggs the natural colour of good ones, by washing them in a peculiar sort of wine, and this practice has sometimes been resorted to by unprincipled dealers to palm off their old and useless seed as good and fertile.

The eggs are laid in August, and as it is of the utmost importance that they should not produce caterpillars till their food is ready for them, i.e., till the mulberry trees have begun to put forth their leaves in the succeeding season, their hatching time ought not to arrive till the following May. But too high a temperature during the early months of the year might produce a premature hatching, which would, of course, be the ruin of all that thus come out. Therefore great pains are taken to shelter the eggs from an excess of heat during the winter. They can stand any amount of cold, and in travelling, are often without damage packed in ice. They need also to be kept dry, or they will become mouldy, and of course perish. A cold, dry cellar, consequently, is about the best place for them during the winter. But they must also be protected from rats, mice, insects, etc., any of which creatures would soon make short work of them. Sometimes they are put in bags, and hung up to the ceiling in a dry, cold room, in which position they are well out of the reach of most of such depredators. They are also kept in tin boxes with perforated sides.

When the mulberry trees begin to show their leaves, preparations are made for hatching the eggs. If they have passed the winter adhering to the sheets of cloth or paper, all that is necessary is simply to spread these out on trays or in boxes in a well-aired and warm room whose temperature averages about 75° F. If they are loose, they must not be placed in piles, but uniformly sifted or spread out on pieces of paper or cloth, preferably the latter, as the young grubs then get a better foothold to help them in working their way out of the eggshell. The temperature should be kept as uniform as possible, and the heat gradually increased about 2° per day, and then, in about a week, the young caterpillars may be expected to begin to appear. There should be good ventilation in the room, but the direct rays of the sun must on no account be allowed to play upon the eggs, or they will speedily be killed. Nor should the atmosphere be too dry, or the eggshells will harden so much that the young caterpillars will have difficulty in getting out of them. In large establishments or magnaneries, as they are called, the floor is occasionally sprinkled with water to produce the requisite degree of moisture. Of course, when the numbers are small, these

elaborate arrangements cannot be made, and recourse must be
had to such substitutes as can be provided; *e.g.*, the eggs, might
be placed on the mantelpiece over the ordinary sitting-room or
kitchen fire, and if there be a difficulty about keeping up the
requisite warmth at night, they might be taken into bed with one,
if their company be not objected to ! With us, however, it will
generally be the case that the eggs hatch before the mulberry
leaves are ready for them, unless precautions are taken to keep
them sufficiently cold. Under such circumstances, they must be
fed with some substitute such as lettuce, which, however, must be
carefully wiped before being given to them, that it may be quite
dry, otherwise it will injure their health.

It will soon become evident, from the changing colour of the
eggs, when the first hatching may be expected. Even at the
very commencement of their larval life, the insects observe those
methodical and punctual habits which never after desert them,
and in the majority of cases they crack their eggshells in the
early morning hours, between three and eight o'clock, and by far
the greater number of the same batch make their *début* on two
successive days. There will be a few who are in a hurry to
commence their larval life, and issue a day or two before the rest.
—" galloping worms," as they are called in China; these may be
allowed to perish for their temerity, for it is very important to
keep all at the same stage of development, in order that they may
perform their moults at the same time. For a similar reason, those
of the second day's grand hatch should be kept separate from those of
the first. Any eggs that have not hatched by the fourth day after
the first caterpillars appeared may be thrown away, as they yield
only inferior and feeble insects, which are not worth the trouble
of rearing.

Now comes the feeding business. The true food of *Bombyx
mori* is the leaves of the white mulberry tree; but in this country
it is chiefly the black mulberry that is cultivated, the leaves of
which are coarser than those of the other species. Where
mulberry leaves cannot be obtained lettuce may be used as a
substitute, but the worms do not thrive so well on it, and the
operator must not be surprised if he loses some of them by death,
and if the cocoons of the rest do not turn out so good as he had
hoped and expected. Other substitutes, such as dandelion,
cherry, black-currant, have been used with more or less success,
but it cannot be too strongly insisted upon that to produce
a thoroughly successful rearing, every effort should be made to
secure mulberry leaves. When the young caterpillars first emerge,
the leaves are still very small and tender, and well-suited to

their juvenile capacities, a fortunate thing for them, for it would be as unsuitable to offer them the hard, rough, full-grown leaves as to feed a baby on beef.

But now a difficulty arises. Here is a set of crawling mites perhaps some hundreds or even thousands in number, scarcely more than one-twelfth of an inch long, amongst a lot of unhatched eggs. How are they to be fed in such a way as not to interfere with the hatching of the rest? They are of course much too small to be picked up by the fingers, and if one were to attempt this, the only result would be that they would be crushed to death. In fact, even when they grow larger, they should not be touched with the hand any more than can possibly be helped. All such unwarrantable interference with their persons is likely to damage them in one way or other and hinder their proper development into a moth. Young people who keep caterpillars of any kind are far too ready to nurse and pull them about, and in moving them from one place to another, to seize them with no very gentle grip between finger and thumb. But any one who thinks at all about the matter, and remembers what was said in the last chapter about the internal organization of the caterpillar, will see at once that such treatment is all too likely to produce internal bruises or other damage that may prove disastrous or even fatal. And apparently young people are not the only ones who sin in this respect, for the Chinese, in their enormously extensive silkworm literature, have numberless warnings on this very point, and some of their peasantry must, at one time or other, have been most careless in the performance of their duties, or we should hardly read such words as these : " Silkworms are very tender things, and cannot bear being rubbed or pushed ; while they are young, people know that they must deal gently with them ; but when large, in separating them or removing their excrements, people sometimes lazily roll them about without any regard ; or leave them for a long time huddled up in confusion, or pitch them high and throw them far, from which causes much injury and sickness arise."

Amateurs in this country usually employ a camel's hair brush to pick them up with. This, if one does not grudge the enormous expenditure of time it involves where the numbers are large, answers pretty well when they are quite young ; but it is not always easy to drop the little caterpillar when once he is on the brush, without working and twisting the latter about a good deal —a process which can hardly be beneficial to the entangled grub. When the caterpillar is fully grown, such a method is inapplicable, and the brush can only be used to give the creatures pokes and pushes till they roll off into the desired place.

The best plan is to leave the caterpillars to transfer themselves to the food. This may be done as follows : place a piece of open netting or gauze such as is shown in Fig. 17 over the newly hatched worms as they lie amongst the eggs, and on that put gently either a few small separate leaves, or chopped pieces of such, or a twig with small leaves on it. The little insects soon become aware of the presence of their food, and pass through the meshes of the gauze to it and at once set to work on the great business of their life. When all have thus clambered on to the food, the netting may be lifted, and the whole transferred to the quarters the insects are permanently to occupy. The unhatched eggs are thus not interfered with, and if another batch comes out the next day, a similar arrangement may be made for removing them. If they are picked off with the brush, they must be deposited on the top of some small whole or chopped leaves previously placed evenly on the bottom of the receptacle in which they are to be kept.

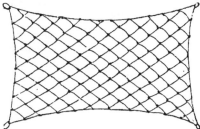

Fig. 17.—Net for removing Silkworms.

In small rearings, the caterpillars are best kept in shallow cardboard boxes or trays, without covers ; but forethought must be exercised as to the number of these to be provided, for, as during their larval life, the worms increase upwards of nine thousand times in weight, they will, of course, require far more room during their later ages than at first, and a set of boxes which would afford them ample accommodation during the first week or two of their life, will at the close of the fifth age be ridiculously inadequate. For large " educations," a room is fitted with shelves, on which large trays can be placed, or the shelves themselves are covered with clean sheets of paper and the worms placed on these. In such cases the shelves have a little ledge at the margins, to prevent the worms falling off. Sometimes large frames or hurdles made of reeds are substituted for the trays. It is well to cover the bottom of each tray with a clean piece of paper, which can be changed each day, and thus prevent the tray from becoming soiled.

Silkworm rearers must not be too fond of sleep, for the caterpillars eat most heartily early in the morning and late at night, and they should always be well supplied with food in a perfectly fresh condition. It must be remembered that the quantity and quality of the silk depend on the quantity and quality of the food the caterpillar gets, and the regularity with which it is supplied. The rearer must be prepared to be assiduous in his attention to his charges during the whole of the five or six weeks they will require his care, and neglect during a single day would probably cost the lives of most of his brood. When the rearing business is carried out in real earnest, the first meal is usually given about five o'clock in the morning, and the last between eleven and twelve at night. While the caterpillars are very small they will need constant attention, and should be fed five or six times a day with young and tender, or with chopped leaves. The chopping of the leaves is to furnish more edges for them to attack, for this is where they always begin munching, holding the food between their legs. Frequent feeding ensures their having the food fresh ; and it is better to renew the food frequently, giving less at a time, than to provide them all at once with a quantity sufficient to last them over a longer time ; the tray is to be regarded rather as a dinner-table on which a meal is spread, than as a larder in which the occupants can help themselves when they please. It is, moreover, a truer economy of food, to give them only sufficient for present needs, than to supply them with enough to last over several meals, when much will unavoidably be wasted through becoming shrivelled and soiled.

Care must be taken to distribute the food uniformly, so that all may get the chance of eating the same amount; for the duration of the " ages " depends upon the amount eaten, and if the supply to some caterpillars be more scanty than to others, they will not all moult together, and then, while some are lying in the torpor of their " sickness," but the others still vigorously eating, and therefore needing to be kept supplied with fresh food, the moulters stand a chance of being damaged by having the leaves lying on their backs, or by their fellows pushing up against them or crawling over them, and may even be thrown away unwittingly during the removal of the litter. If they are all moulting at the same time, no such difficulty occurs. It may seem to some that this keeping of the worms all in the same stage of development is a trivial matter to lay such stress upon, but in reality it is very important, and much inconvenience will be saved, as well as probable damage to and loss of caterpillars prevented, if it be attended to.

Even the gathering of the leaves needs care and forethought. They should never be given to the insects wet, as in that case they would affect their health. Therefore, if the leaves for the early morning meal be picked just before the meal, when they will most likely be wet with dew, they must be carefully dried before being distributed to the worms. Indeed, it is better to pick the leaves for this meal over night, and keep them during the the night in a cool dry place; in this way they will be much more wholesome. Considerable difficulties will, of course, be experienced in showery weather, and advantage must be taken of the intervals of sunshine to gather such food as may be necessary, so that the stock may not come to an end before the next opportunity occurs of filling the larder. If a storm threaten, a considerable stock of leaves should be gathered, sufficient to more than last out the probable duration of the bad weather.

A new meal is given by simply laying the leaves carefully on the worms, distributing them evenly so that all may share alike. But it is necessary, at least once a day, to remove the fragments of the repasts and the excrement; when this is done, a net with meshes large enough for the caterpillars to creep through, or a piece of similarly perforated paper, should first be laid across the tray, and then a supply of leaves on this; then, when the worms have crept through to the fresh food, they may be removed to a duplicate tray already provided for their reception, or simply set aside while the old one is being thoroughly cleansed. Of course two nets will be required, for the first will remain beneath the worms till the next meal is supplied, when the second is placed over the tray with the fresh leaves on it, and the worms clamber through again, thus leaving the first net free to be taken out, cleaned and reserved for use at the next meal. Of late years perforated paper has been largely used instead of the nets, a peculiar kind being specially manufactured for the purpose.

Cleanliness is of the utmost importance, especially during the later ages, when the excrement, or "frass," as it is termed, accumulates with alarming rapidity, and if allowed to remain long in the trays, it is apt to become mouldy and disagreeable, and to soil the leaves and engender disease. Therefore during the last age, it is well to use the net at each feeding, so that the trays may be kept clean and fresh. Ventilation, too, is another very important matter. The insects should always have plenty of fresh, pure air, but care must be taken that, in providing this, they be not exposed to draughts, nor be allowed to become chilled; a uniform, moderately warm temperature is of great importance to them, and a sudden advent of cold weather will, if they be not well guarded against it by artificial warmth and shelter, retard their development, or may even prove altogether fatal. As an example of the disastrous effects of a chill, a story is told of a cultivator who had successfully reared 30,000 worms, up to the point when they were just ready to spin. At that very time, there came a chilly north-east wind, and, in consequence, many of that large brood assumed the pupa state without forming a cocoon at all, and the great purpose for which they had been reared was frustrated. On examining these, it was found that the cold had had the effect of congealing the gummy substance of which the silk should have been made, and therefore they had not been able to cause it to issue from the spinneret.

Some of the Chinese instructions on the subject of cleanliness and ventilation are amusing, as the following specimens will show. " Dirty people must not come into the room, and it must be kept free from filth. Let no leather nor hair be burnt near them. They must not be brought in contact with the fumes of a frying pan, or the smell of coal. They dislike wine, vinegar, and all kinds of strong tastes, the smell of raw meat and fish, musk, etc. Don't carry scent about with you. Let not any one eat ginger or broad beans in the room. In the winter months, collect a heap of cow-dung, which, on the approach of spring, may be worked up into cakes and dried; these when burnt emit an odour that is agreeable to the worms." Still more manifestly savouring of superstition and notions of ill-luck are the following. "Strange persons must not enter the room, nor persons in mourning. They cannot bear the sound of crying and weeping in the room, also filthy and wanton conversation ! "

We will now suppose that the silkworms have been successfully cared for during the first five or six days of their life. The first moult may now be expected. Signs of it soon appear in the failing appetite of the insects. Feeding may now cease, for when once the "sickness" begins they will eat no more till they have changed their skins. No more meals, therefore, are given, but the trays are cleansed as much as possible, so that the little creatures may have everything in their favour in the trying period that awaits them. The duration of this "sickness" is from twelve to twenty-four hours, some performing the moult more quickly than others. It is best to wait till the majority have changed, before giving the first meal of the second age, in order that the whole batch may be kept well together. Any laggards that have not changed by the time the majority are ready may either be removed to be separately tended, or destroyed, if there are plenty of others, for they will be a good deal of trouble, and most probably will not yield a sufficiently good return in silk to justify the pains expended on them.

When all are ready, the eating may be allowed to begin again, and a bountiful supply given, as they will be extremely hungry after their long fast. During the second age the feedings go on as before, the caterpillars being allowed more and more room as they grow, and the quantity of food being also gradually increased. The second and third moults are much like the first, and most will probably get successfully through them. By this time, however, it will have been found necessary, in consequence of the rapidly increasing size of the worms, to divide the batch. This can easily be done by aid of the net. If, after this has been put on with a meal, it be removed when only half the caterpillars have ascended to it, these can be transferred to a new box, and the rest cleaned as before. During the third and fourth ages about four meals a day will be required.

The fourth moult, which introduces the last age, is accomplished with much more difficulty, and apparently with much more discomfort, and most likely several will perish at this stage. This "sickness" lasts from thirty-six to forty-eight hours, and during its progress there will be perceived a sickly and disagreeable smell about even the cleanest trays. After this moult it is best to give the forward ones a feed as soon as they are ready, without waiting for the rest, for, as there are to be no more moultings, there is not so much reason for keeping the batch together, and besides this, the length and quality of the silk in the cocoon depends very largely upon the amount of food taken during the last age. Therefore let them have as much as they can eat. The amount of leaves required during this last age is something enormous, and in a large "education," the attendant will be kept fully occupied in supplying the voracious appetites of the worms. This is especially the case during the last two or three days before spinning, which is the period of greatest excitement and rush on the part of the attendants. During these few days, as already mentioned, about half of the entire quantity of food taken during the whole larval existence is consumed. Therefore feedings and cleansings must be very frequent, and the greatest care should be taken to remove at once any that appear diseased, or they may soon infect the rest, and cause great mortality. This is the chief time for diseases to manifest themselves, and the attendant must therefore be continually on the watch for them.

At the close of the fifth age comes the "mounting" or "ascending" season. The silkworm ceases to eat, becomes transparent like a grape, and begins to emit silken threads, manifesting, at the same time, a certain restlessness, and climbing *upwards* wherever it can. It also shrinks somewhat in size, and voids most of the excrement it contains, not as a solid mass as heretofore, but in the fluid form. It must on no account be permitted to form its cocoon amongst the food, as the silk will be discoloured by the bits of leaves and fluid excrement, and will therefore become useless. Warned, therefore, by the above signs, the amateur has provided a quantity of little conical paper bags, of such a shape as grocers use for wrapping up ounces of mustard, pepper, etc., and of such a size as a half sheet of note paper will make. Each caterpillar, as it manifests the wandering tendency, is placed in one of these, and the bag is then pinned up to the wall. It need not be closed, as the worm will not escape, but begin at once to form its cocoon. This method of course necessitates the taking up of each worm separately as its turn comes, putting it in the paper bag and pinning it up, and therefore involves so considerable an expenditure of time that it would be an altogether inapplicable mode of procedure in connection with a large magnanery. In such a case, therefore, as with the feeding, the worms themselves are made to effect the needful transference, and thus a vast amount of labour is saved.

For this purpose, bunches of heather are set upright at regular intervals along the shelves, with their tops bending over towards one another, so as to make a series of arches (Fig. 18). These are to serve as a lodgment for the cocoons. The caterpillars soon find their way into the heather, and at once begin to construct their silken nests in the interstices between the branches,

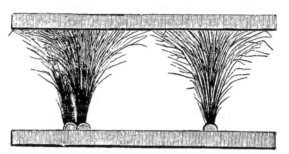

Fig. 18.—Heather twigs to receive cocoons.

in this way the heather in a few days becomes crowded with cocoons as with a crop of fruit. Sometimes two caterpillars enter into partnership over a single cocoon, which, as it necessarily consists of two separate threads which cross one another in all manner of ways, and therefore interfere with one another on unwinding, is useless for commercial purposes. They are called "double cocoons," and of course contain two chrysalises; they are much rounder than the normal form. If, therefore, any caterpillars be observed thus uniting their efforts, they should be separated and placed on different parts of the heather, or their combined labours will only be so much lost energy. Of course double cocoons are never made when the caterpillars are placed in paper bags. Occasionally, even more than two combine to form a single cocoon, which thus envelopes them all.

During the process of spinning, the temperature ought to be kept tolerably uniform, and should not fall below 80° F.

The cocoons must not be interfered with till the caterpillars have changed to pupæ, and these have become hardened. Not until eight days after the commencement of spinning, therefore, should they be gathered. Then they may be removed from the paper bags, or picked off the heather sprigs. It will most probably be found in all large rearings, that some cocoons are stained with black spots; in these, either the insect has been accidentally crushed, or has died and become putrid, and so spoilt the silk. All such must be very carefully removed, or they will speedily damage the rest. The cocoons ought to be fairly hard to the touch, and able to resist considerable pressure, and any which seem at all soft and yielding should at once be set aside, for they stand a good chance of being crushed, and then if allowed to remain with the rest, would certainly stain and so ruin them.

After these needful separations have been made, the "floss" may be torn off and set aside, and if the cocoons are numerous, they may be divided into groups according to colour, size, etc., preparatory to reeling. In extensive "educations" the number of cocoons is far too great to permit of their being reeled during the few days that elapse before the moths may be expected to issue; therefore, in such cases, the insects have to be killed, lest the moth should pierce the cocoon before it has been reeled. Consequently if it be contemplated to continue the breed the next season, some cocoons must be selected for breeding purposes, before this great slaughter takes place, and if it be really desired to obtain the best possible results, the very finest cocoons should be selected for the purpose. The silk of these will be sacrificed, for they are to be set aside without being unwound, in order that the insects may complete their development, and then the moths, in issuing from their confinement, will pierce the cocoons, which will thus be rendered unfit for the market. But where only a very small number are concerned, there is not such necessity to kill the chrysalises, as there will probably be plenty of time to reel all the cocoons before the moths appear. Even in such cases, however, it is best to select for breeding purposes some whose silk may be sacrificed, in order to provide against the possible failure of the breed through any damage to the pupæ during the process of reeling.

In selecting for breeding purposes, of course an equal number of males and females should be saved. But here comes a little difficulty; there has been no such thing as a distinction of sex up to this point; how then are we to tell which cocoons will produce male moths, and which females? It has been generally supposed that the male cocoons are smaller and less rounded at the ends than the females; but though it is no doubt generally true that the largest cocoons will produce females, it is impossible to be absolutely certain, and therefore rather more should be saved than are theoretically necessary, in case the educator's discrimination should have been much at fault. The magnitude of the projected breeding of the next year will of course determine how many insects' lives are to be spared. Reckoning that each female will produce about three hundred eggs, we should require, for example, if an "education" of thirty thousand be projected, two hundred moths, half females and half males. Therefore some two hundred and twenty or two hundred and thirty cocoons ought to be saved; and for smaller numbers in the same proportion. The best for breeding purposes are not necessarily the largest, but those that are the firmest to the touch and the most delicately coloured, except in the case of the green cocoons, when the greener they are the better. If there are any double or treble cocoons, these may be used for breeding purposes, though they will be useless for reeling.

The breeding cocoons are pasted close together, side by side, on sheets of cardboard, in rows which are at the distance of a little less than a moth's length from one another (Fig. 19). The cardboard is then placed in an inclined position so that each moth as it comes out may easily find a resting-place on the

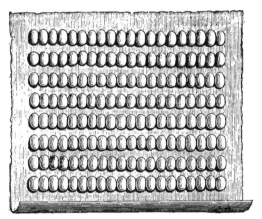

Fig. 19.

cocoon immediately above it, and cling to it in a suitable position to dry its wings. Another method of arranging them is to thread them on a string like birds' eggs, taking care, however, on inserting the needle, to pass it only through the silken covering, and not to injure the chrysalis within. The strings may then be hung up like strings of onions. The male and female cocoons should be put, as far as possible, on different sheets of cardboard, an attempt at their separation being made as above, or by weighing the whole lot so as to calculate the average weight of a single cocoon, and then considering as females those above the average, and as males those that fall below it.

These arrangements having been completed, the rest of the pupæ, which will form by far the larger proportion of the whole stock, must be killed. This is done by placing the cocoons in wicker baskets in an apparatus where they are exposed for about twenty minutes to the steam of boiling water. Fig. 20 shows a section of such an apparatus.

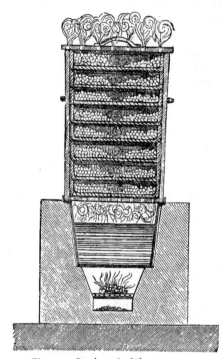

Below is seen a small furnace which heats the water in the vessel just above it, the steam from which passes through the trays containing the cocoons, and issues at jets at the top. When thus killed they must be thoroughly dried first by exposure to the sun and then by being strewn on wooden shelves in the shade, where they can have plenty of air. They are occasionally stirred to give all an equal chance, and then after a couple of months they will be quite dry and may be kept any length of time, provided they are preserved from the attacks of rats, mice, and beetles. Cocoons in which the pupa has been killed are said to be "choked," while those in which it is still alive are

Fig. 20.—Section of stifling apparatus.

called "green." This, of course, has nothing to do with the colour of the cocoon.

Amateurs usually do not take the trouble to separate breeding cocoons, but reel all they have, until only a thin shell of the cocoon is left covering the pupa; they are then placed, with or without the remains of the cocoon, on a bed of bran in a box, where they are kept till the moths issue.

In about a fortnight or three weeks from the time of commencing the cocoon, the moths will begin to appear. Like the caterpillars, they make their entry into the world in the early morning, from four to eight o'clock. As they appear, they should be seized by the wings and placed in boxes, so as to keep the sexes apart for a time. The males may be known by the incessant fluttering of their wings ; the females are much quieter. When a sufficient number have been collected, the sexes are to be placed in equal numbers on sheets of paper, cardboard, or cloth, and kept in the dark. They will soon couple, and may be left to themselves for at least half a day. At the end of this time, the females should be placed on blotting paper for a short time, till they have discharged a greenish yellow fluid. They may then be transferred to the sheets of cloth or paper, which are to receive the eggs, when they will commence laying at once (Fig. 21). The sheets may be placed smoothly in trays, and the moths set in rows upon them; darkness is still essential, but, at the same time, plenty of fresh air should be allowed to reach them. All the best eggs will be laid during the first twenty - four hours ; any deposited after

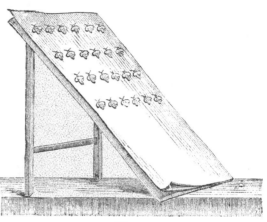

Fig. 21.

that time should be kept separate from the rest, as an inferior stock. When the eggs have all been laid, the only business of the moths, adult life is finished and therefore in a few days they die.

We must now return to the cocoons that have been put aside for reeling. In small rearings all are usually reeled, and in such cases they are not first " choked." They are placed in a basin of warm water, and moved about in order to soften the gummy substance which binds the coils of thread together, and to enable it to be unwound. Then the end of the thread is sought for and gently pulled. If the water is at the right temperature, the thread will unwind readily and evenly; if it comes off in masses, the water is too hot, and causes the separation, not of the simple thread, but of a whole patch of the double curves spoken of in a former chapter; on the other hand, if the cocoon does not readily unwind, but is continually being pulled out of the water, through the thread adhering too tightly, the temperature is not high enough. Amateurs usually reel only a single thread at a time, but the silk so prepared is useless for manufacturing purposes, the staple of commerce always consisting of several threads combined.

The amateur's reeling machine is a very simple affair. It

merely consists of a sort of wheel something like the paddle wheel of a steamboat. It may easily be made for oneself. First, take two flat pieces of wood about eight inches long, and a quarter of an inch wide, and fasten them upon one another in such a way that they cross at their middle points, and are at right angles. Then make another similar cross. A piece of stout, strong wire, about six inches long may now be run through the centres of the two crosses, which are to be firmly fastened to it, in such a way that they remain parallel to one another, and about three inches apart. They will thus form the spokes of a double wheel. Next, make four small, flat, smooth pieces of wood of similar breadth to the spokes, but only about half as long, and fasten each of these across, from the end of a spoke of one cross to the corresponding one of the other, and the wheel is then complete. A good way of fastening the cross pieces to the spokes is to taper the ends of the latter to a cylindrical form, and then insert them into holes made near the ends of the cross pieces. One of the latter should be fastened in such a way that it can, when required, be slipped farther down the spoke. The wheel will now need two supports on which to turn; these may be two upright pieces of wood, each with a perforation near the top, and fastened to a board in such a way that the axle of the wheel may be run through the perforations. Then a handle may be attached to one end, and the whole is in working order.

The end of the thread of the cocoon is now fastened to one of the cross pieces of the wheel, and the latter is turned, when the thread will be wound off the cocoon, which still remains in the warm water, on to the wheel. The basin must not be placed too near the reeling machine, or the silk as wound will be too wet and the threads will stick together. The cocoon should be guided by the unoccupied hand of the reeler, in such a way that the thread does not always occupy the same position on the wheel, but is shifted backwards and forwards from one side to the other. This is to prevent the different coils of thread from adhering, as they might do if they were always wound in the same line. When no more silk can be obtained from one cocoon, another of the same colour may be begun in the same way, and thus several reeled before removing the silk, in order to obtain a larger skein. When the reeling is finished the movable cross piece may be slipped a little way down its spoke, when the coil of silk will, of course, become slack, and may be taken off the wheel after the latter has been removed from its supports. The ring of silk thus obtained may be drawn out into a loop and tied in a knot in the middle, in the same way as a skein of thread. As a saving of time, several cocoons may be reeled together, if one does not object to the threads becoming intermixed in the skein.

On the large scale the reeling is performed by women, and the establishment in which it is carried out is called a "filature." Each operator has in front of her a metal basin of water, the temperature of which can be regulated by jets of steam allowed to enter it through taps under the operator's control. The thread is never wound singly, as it is much too fine for manufacturing purposes but those of several cocoons are twisted together and combined into a single thread. A common number is five. The woman puts a number of cocoons in the basin and moves them about so that the gum which fastens the threads may become uniformly softened ; then with a little brush made of birch twigs split at their ends, she gently beats them as they lie in the water, till the threads of all have become entangled in the brush. She then takes the bundle of threads and shakes it till each cocoon is seen to hang by a single thread. Then selecting five she unites them, and passes the combined thread through a little glass eyelet fastened near the side of the basin. Then taking another set of five she does the same with them, passing them through another eyelet placed on the other side. The two strands thus made are then brought together above the eyes, crossed and twisted together several times, and then again separated above the twist and introduced into two other glass eyes, through which they are led, one to each end of the reel. This is much larger than the amateur's contrivance mentioned above, and usually has six spokes instead of four. It is kept revolving steadily and rapidly, and at the same time moved a little alternately from side to side, so that the threads may be crossed and not stick together.

It is very important that the thread should be of uniform diameter throughout, and that the different elements of which it is composed should be well twisted together and form a compact strand. It must also be completely freed from moisture. These objects are attained in some degree by means of the eyelet holes, the twisting, and the lateral movement of the reel, together with a suitably regulated distance between the basin and the reel. The uniformity of the thread, however, depends a good deal upon the watchfulness and skill of the operator. For the threads of all the cocoons will not necessarily be of the same length, and some may give out before others are finished. In such a case, a new thread must at once be introduced, or the strand will become of an inferior character. The silk, too, becomes finer as the end of the cocoon is reached, for, in the spinning, the silk glands of the caterpillar were then almost exhausted ; therefore as the end of a set of cocoons is reached, it is necessary to introduce additional threads to keep up the uniformity of the strand. The operation of introducing new threads is called "nourishing" the silk, and it is done by dexterously casting a new thread on the combined strand, to which it instantly adheres. It is a difficult operation and requires considerable practice.

The operator has a perforated skimmer to remove the pupæ as the cocoons are finished, together with the refuse silk; and also, by her side, a basin of cold water, to cool her fingers occasionally.

After the reeling, the staple is "cleansed" by passing it through a clasp lined with cloth, by which any loose silk is removed; it is then twisted about 500 times to the yard, then doubled and twisted again 400 times to the yard, after which it is run on to a comparatively small reel, about 18 inches in diameter, from which it is ultimately taken off and twisted into a peculiar knot or hank. From the above, it is evident that the staple, as finally prepared, consists of at least ten of the threads as originally formed by the silkworm.

The floss, together with the refuse of the cocoons, cannot be reeled, but, after being suitably cleansed, is torn up or "carded," and then spun in the same way as cotton, thus making an inferior quality, which is called "floss silk." There are two other qualities of raw silk, viz., "organzine," made from the best cocoons, and much twisted, and "tram," made from inferior cocoons and only a little twisted. These are the forms in which it is handed over to the manufacturers of silk goods, and from which, when suitably dyed, they work up those magnificent fabrics which are amongst the most costly and splendid of all the textile materials used for decorative purposes.

Notwithstanding the extraordinary length of silken thread that can be uncoiled from a single cocoon, it yet takes an incredible number of insects to yield even a comparatively small weight of silk. It may generally be reckoned that little short of 3,000 cocoons will be required to yield a single pound of silk. Now, it has been estimated that the annual production of raw silk in the Chinese Empire alone is about twenty-four million pounds; reckoning, then, at the rate of only 2,000 cocoons to the pound of silk, this would imply the rearing of the utterly inconceivable number of about fifty billions of insects annually. This is probably not an exaggerated estimate, for the 3,500 tons before mentioned of European-bred raw silk annually presented to the markets involves the rearing of at least twenty billions of silkworms, and China is a far larger producer than all Europe put together.

All that has been said hitherto has had reference only to the

Common Mulberry Silkworm, *Bombyx mori.* But this is not the only Bombyx which feeds on the mulberry and is domesticated. In India several other species are reared, all mulberry feeders, which were for a long time confounded with *B. mori.* Of these, however, none produce such large cocoons nor such good silk as the common species. They differ from it, too, in the number of broods produced in a year. *B. mori* always produces one brood only, and is therefore called an "annual;" most of the other species produce two, three, four, six, or even eight broods per annum, and in the commercial world are, for that reason, distinguished as "bivoltins," "trevoltins," "quadrivoltins," etc. In the case of bivoltins, the eggs of the first brood hatch very soon after being laid, to produce the second, the eggs of which remain through the winter and yield the first brood of the following year. When more than two broods are produced in succession in the same year, the life of each brood extends over a shorter period, and the number of moults is reduced to three instead of four. Notwithstanding the large number of broods produced yearly by some of the "polyvoltins," the "annuals" are found to be the most profitable, and are therefore much the more extensively reared.

Bombyx textor, a species domesticated in S. China and Bengal, is, like *B. mori,* an annual; it produces usually a white cocoon, more flossy than - that of the common species. *B. cræsi,* the Madrassee, is cultivated in Bengal, having been introduced from China. It yields seven or eight broods in the year, and forms golden yellow cocoons. *B. fortunatus,* the Dasee of Bengal is also a polyvoltin, and is noted as making the smallest cocoon of all. The silk is yellow. *B. Arracanensis,* the Burmese silkworm, is domesticated in Arracan, whither it is said to have been introduced from China; it spins a beautiful yellow or white silk, not quite so glossy as that of *B. mori,* and yields several broods annually. All these species are domesticated mulberry feeders.

CHAPTER V.

THE SILKWORM—ITS DISEASES AND IMPERFECTIONS.

SILKWORM rearers usually reckon that, granted good "seed," there are five requisites to success in their employment, especially if large numbers are operated on. These are, first, uniformity of age in the occupants of the same tray, so that all may moult at the same time; second, no intermission in the supply of food, except during the moulting sicknesses; third, plenty of room; fourth, good ventilation, and a plentiful supply of fresh, dry air, at a uniform temperature, especially during the fourth and fifth ages; and lastly, cleanliness, which becomes of increasing importance with the advancing age of the larvæ. But even if all these conditions are strictly fulfilled, the silkworm rearer must be prepared for disappointments, and must not expect to bring all his brood successfully to their perfect state. There are many causes of loss and imperfection, and it is to these that we must now turn.

A little difficulty is occasionally experienced by the insect, even in making its exit from the egg. Sometimes the hole through which it is to escape is not quite large enough to allow the passage of the head, and the young caterpillar is induced to try entering the world stern foremost. This is a stupid proceeding, because of course the head will have to come through sooner or later, if the egg is to be really and effectively hatched. When all the body has issued, then comes the attempt, of course a futile one, to drag the large head through, and the creature becomes stuck in that position, wearing the old eggshell as a bonnet, and necessarily becoming soon starved to death.

But one of the commonest sources of loss will be the moulting sicknesses. Changing the skin, especially at the last two moults, seems to be a very trying operation, and in a large brood there are sure to be some who will succumb to the severity of the trial. Such are called *luzettes*. There is no remedy for this disaster, and the luzettes must simply be left to their fate, and removed as soon as it becomes evident that they will not spin; for they will speedily die, and their corpses will contaminate the air. If a caterpillar receives a wound, the scab formed on its healing will more or less interfere with the next moult; the old skin will adhere at the spot, and the insect, unless it receives external help, will either be unable to remove it beyond that point, or will at least be considerably inconvenienced by it, and will probably either die or become more or less of a cripple.

Unwholesome or too juicy food produces diarrhœa, the remedy for which is, of course, the substitution of a more solid diet. Too great heat, again, renders the worms feeble and sickly; and sometimes they become yellow and limp, and soon die of a sort of jaundice, called *grasserie*. Sometimes, again, a larva just manages to get through its moult, but exhausts all its energy in the process, and has not enough left to recommence eating, and therefore perishes. Such are called *arpians*.

There are, however, three diseases which far exceed all others in importance, and which have on different occasions, caused fearful mortality amongst the insects. They are known by French names as *Flacherie, Muscardine,* and *Pébrine.* Flacherie, or flaquerie, is a disagreeable and particularly disappointing disease, as it attacks the caterpillars just before their change into a chrysalis, and very rapidly produces fatal results, thus rendering useless all the expense and trouble that have been incurred in bringing them up to that point. The caterpillar feeds up well, eats a great deal and grows very fât, but when it is nearly full fed and its rearer is beginning to look forward to some reward for his pains, it suddenly, while still appearing healthy, becomes exceedingly inactive, stretches itself out on its food and remains without moving. It becomes limp and flaccid, and so rapid is the spread of the disease, that in twenty-four hours the contents of its body turn black and putrefy. This disease may be produced by overcrowding, and defective ventilation, by the presence of an undue amount of moisture in the air, or by the use of food which is damp, or too succulent. As soon as any caterpillars are seen to be affected they should be removed, or their putrefying corpses will contaminate the rest. It is possible for flacherie to become hereditary; or it may be developed suddenly in a brood and kill off numbers of them at once. A silkworm which has died of this disease is called in French a *mort-flat.*

Muscardine is a disease of quite a different character. It attacks the caterpillars in all their ages, but is by far the most fatal during the period between the last moult and the formation of the cocoon. It is impossible to detect when the insect becomes smitten by the disease, for, outwardly, it at first appears perfectly healthy, though it may have the germs of the disease within. When about to die, however, it becomes languid, and its dorsal vessel pulsates more feebly. It soon dies, and after a while becomes reddish in colour and perfectly stiff and rigid. About twenty-four hours after this, there suddenly appears a fine white powder covering the body. This disease is either caused, or, at any rate, accompanied, by the development of a minute fungus, called *Botrytis Bassiana,* in the insect's body, and the white powder above referred to consists of its spores. A somewhat similar disease attacking house flies is probably familiar to every one. Towards the close of autumn we sometimes see the flies dead, though in the attitude of life, and adhering to the window panes, while around them on the glass is a sort of white halo, which consists of the spores of the fungus that has established itself in the body of the fly and spread so much as to penetrate all its tissues and kill it. This is not the same species as is the cause of the disease muscardine, but it is of a similar nature.

Muscardine is a contagious disease, *i.e.*, the spores of the fungus, on gaining access to the body of a healthy caterpillar, as, for example, by passing in at the spiracles, will generate the disease ; or even if they settle on the skin they are able to produce fine threads which can penetrate the tissues of the living caterpillar, and so work its ruin. If, therefore, any specimens in a brood become infested with muscardine, it is of importance, not only that they should themselves be removed, but also that the trays and other apparatus should be disinfected, as, for example, by washing with a dilute solution of carbolic acid. There seems to be no cure known for this disease, but the best means of prevention are, as usual, good food, a plentiful supply of pure, dry air, and cleanliness. About fifty years ago great damage was done to the French silk industry by this disease, but thanks to scientific investigations into its origin and cause, and to improved sanitary appliances, it has now almost entirely disappeared. The muscardined caterpillar, though it usually perishes before forming its cocoon, is yet sometimes able to accomplish this task, but the cocoon so produced is not a good one ; it has a sort of waxy semi-transparency, and contains, not a living chrysalis, but the dead and dried-up caterpillar.

The most terrible of the diseases of the silkworm is *pébrine*, or pepper disease, which was formerly called *gattine*. It has acquired the name of pébrine from the fact that the caterpillars affected with it become covered with a number of black spots, as though they had been sprinkled over with coarsely-powdered black pepper. It is a disease something like cholera, but possesses the terrible distinction of being hereditary. This, of course, in the case of an animal which is, and can only be, bred from one generation to another, and of which there is no wild stock to fall back upon, infinitely increases its disastrous effects. Languor, loss of appetite, and the dark spots are the symptoms of the disease. Its cause is to be found in the presence of extremely minute corpuscles, called *psorospermiæ*. These were first noticed by the French naturalist, Guérin-Méneville, in 1849, but it was not then known for certain what was their relation to the disease, and for a long time it was a disputed point whether they were the cause or merely an effect. It remained for the renowned Louis Pasteur, now so well-known by his experiments in connection with hydrophobia, not merely to demonstrate that the disease is entirely dependent upon the presence and multiplication of these corpuscles, but also to show how it might be stamped out.

The corpuscles are excessively minute, shining, oval bodies, so small as to require the highest powers of the microscope for their examination. They fill the silk tubes of the infected insect, which may then go through the movements of spinning but without being able to produce any thread; they enter its digestive canal, and spread throughout its body, and are not confined to any one stage of its existence, but appear equally in the egg, larva, pupa, and perfect insect. Wherever they are found, there the disease becomes developed, and where they do not occur, the disease is equally absent. For these reasons, it is not merely hereditary, but also contagious and infectious. For, as some of the corpuscles are sure to be passed with the excrement of a diseased individual, any leaves that happen to become soiled with this, and are then eaten by a healthy individual, will serve to convey into its body the insidious foe. Or, again, if a healthy individual should be bitten or scratched by a diseased one, the minute corpuscles may in this manner be introduced into its system, and infect it. In these ways, of course, a single individual may infect a whole brood. Pébrined silkworms may die during any of their ages, though here, as with muscardine, the greater number perish in the last. This applies only, however, when the disease is inherited. If, on the other hand, a healthy caterpillar should become infected by contact with diseased ones, or by germs in the dust of the room, it will usually complete its metamorphoses, and may even form an excellent cocoon; but it will inevitably transmit the seeds of death to the next generation, and ruin its posterity. This accounts for the fact, so unaccountable before Pasteur's experiments, that eggs selected from moths that had issued from the finest cocoons, frequently yielded caterpillars that soon developed the disease and perished without yielding a crop of silk.

It was in the year 1865 that Pasteur, at the earnest solicitations of his friend, the celebrated chemist Dumas, undertook to investigate the disease. It had then been raging in France for upwards of twelve years; the silkworms had sickened and died in multitudes, every year increasing the mortality, and the silk industry was in consequence, brought to the verge of ruin. The weight of cocoons produced in 1853 was twenty-six millions of kilogrammes, producing a revenue of one hundred and thirty millions of francs. During the preceding twenty years, the revenue had doubled, and there seemed every reason to believe that future years would see a further augmentation. Then came the pébrine, and by the time Pasteur commenced operations, the annual yield had become reduced to four million kilogrammes of cocoons, a return which, for that year alone, represents a loss of at least a hundred millions of francs. The question, therefore, was really one of national importance; remedy after remedy had been tried in vain; there was no end to the nostrums that were, one after the other, confidently announced as the infallible cure, only to be found on trial absolutely valueless. "Gases, liquids, and solids," writes a French authority in 1860, "have been laid under contribution. From chlorine to sulphurous acid, from nitric acid to rum, from sugar to sulphate of quinine—all has been invoked in behalf of this unhappy insect." But all this was proceeding in the dark; what was needed was a thoroughly scientific investigation of the whole question, to determine the obscure cause and mode of origin of the disease, and then there might be some hope of successfully battling with it. It is no wonder, therefore, that Dumas, who saw in Pasteur just the man for such an investigation, urged it upon him with all the energy in his power, until, notwithstanding Pasteur's attempt to excuse himself on the ground that he knew nothing at all about silkworms, he at length prevailed upon him to undertake the task.

It had already been made out that the corpuscles might appear even in the egg, and it had been supposed that if eggs did not reveal any corpuscles when microscopically examined, they were therefore free from infection and their incubation might be undertaken with every hope of a successful result. But this test, for some reason not understood at the time, often deceived those who put trust in it, and was never widely adopted. Pasteur, however, soon proved that it was quite possible for the egg, and even the worm itself, to be diseased, without the corpuscles being rendered visible by the microscope, and that therefore it was not at all safe to argue the healthiness of a batch of eggs from the absence of visible corpuscles. But he noticed also that the corpuscles increased in size with the growth of their host, being larger in the chrysalis than in the caterpillar, and largest of all in the moth. He therefore determined to deal with the moths rather than with the eggs, in his endeavour to stamp out the disease.

With the utmost care and precision he followed the different steps in the development of the corpuscles. He obtained a large series of perfectly healthy caterpillars from eggs of moths which, by microscopical examination, had been proved to be free from taint. From these he selected sets of twenty or thirty to experiment with. He introduced the disease into the trial set by giving them a meal impregnated with the corpuscles. Taking a small caterpillar already suffering from the disease, he crushed it and pounded up its body with water, and then smeared a little of the mixture over the mulberry leaves. The silkworms ate the leaves, and after a time showed symptoms of the disease ; by killing and microscopically examining different members of the group at intervals, he was able to trace the development of the corpuscles.

The details of one of these experiments may serve to show the care with which the problem was attacked, and at last worked out. On the 16th April, 1868, he infected thirty silkworms by a corpusculous meal, as above. No signs of disease were manifest during the first week, and even on the 25th they seemed well, but the tell-tale corpuscles were detected in the intestine of two, which were then examined. From this time the progress of the disease became rapid, for two which were dissected on the 27th showed the corpuscles not only in the intestine, but numbers of them in the silk glands as well. The next day all the rest showed the black spots characteristic of pébrine. Meanwhile, another set, which had been selected at the same time and from the same stock, and had received an equal amount of food, but had not been infected, and had been kept with scrupulous care apart from the stricken set, grew rapidly and prospered, showing no signs of pébrine, and being by the 30th of April half as large again as those that had contracted the disease. On the 2nd May a diseased caterpillar that had just moulted for the fourth time was examined. Its whole body was found to be crowded with corpuscles, so that it was a marvel it had not already perished. As the disease advanced the worms died, one after the other, till only six of the original thirty were left. These were then submitted to microscopic examination and every one was found to contain corpuscles. This shows the dreadfully contagious character of the disease. They had had only one corpusculous meal, and yet it had been sufficient to infect them all ; not one had escaped. The healthy lot, on the other hand, went through all their transformations successfully, and formed fine cocoons, and the moths, on being examined, were all found to be perfectly healthy except two, which had no doubt become infected during their larvahood by accident.

By such experiments as these it was conclusively proved, that, if the pébrine was to be eradicated, not only would it be necessary to take the utmost care during the progress of any education to guard the silkworms from the possibility of infection, but that only such eggs ought to be reared as had been laid by females which were absolutely free from taint. Pasteur's proposed method of eliminating the disease, therefore, was to allow the moths to lay their eggs, care being taken to identify each mother with her own progeny, and then to kill the moths, crush their bodies, and examine the contents under the microscope, and ruthlessly to burn all eggs produced by such moths as showed the slightest trace of the deadly pébrine corpuscles.

But these proposals met at first with a good deal of opposition. In particular, the cultivators seemed loth to believe that moths apparently healthy, which had issued from good cocoons, could possibly lay any but healthy eggs, whereas Pasteur had discovered, as before stated, that when the disease is contracted during larvahood, the insect will generally complete its metamorphoses and often form an excellent cocoon, but will infallibly produce an enfeebled progeny in the next generation, which will perish before reaching the close of larvahood, and therefore yield no silk. So satisfied was he of the truth of his conclusions, that in order to gain acceptance for his ideas, he even ventured to turn prophet on the subject. Selecting from a certain district fourteen parcels of eggs which had been reserved for hatching, he examined a sufficient number of moths which had produced them, and wrote out his predictions of what would occur to each batch the next season, and placing the document in a sealed packet, he sent it to the mayor of the place, with the request that it should not be opened till the results of the next season's educations had been determined. When the time arrived, the cultivators informed the mayor of the results in each case. Then the seal was broken, and Pasteur's predictions produced, and it was found that in no less than twelve out of the fourteen packets, the issues had been exactly what the celebrated scientist had foretold. There had been many total failures, every single caterpillar in the education having perished; in other cases, nearly all had died, and both these results were exactly in accordance with the prophecy. The two which falsified the predictions had succeeded better than was anticipated, and instead of utterly perishing had yielded half an average crop. Still this was a result that was scarcely worth the trouble of rearing them. It must be remembered that these packets of eggs had all been supposed by their owners to be thoroughly healthy, and had been hatched with the full expectation that they would yield a crop of silk such as would well repay the trouble expended upon them; and yet in almost every case the labour had proved to be in vain. If, therefore, the proposed moth test of Pasteur had really been adopted in this instance, all the batches would have been condemned, and much trouble, expense, and disappointment saved. The above is only one instance out of several, of prophecies both with regard to diseased and to healthy insects being verified. Hence, finally, after much opposition, the method was adopted and proved signally successful, and the silk industry, one of the most important occupations of the peasantry of France, was rescued from ruin.

The brief account given above of the epidemic diseases to which silkworms are liable will suffice to show that the breeders of these insects on a large scale need to exercise great watchfulness, and that the occupation is not by any means the mere amusement it might on first thoughts be imagined to be, but that, on the contrary, it is sometimes an exceedingly anxious undertaking, which may result even in considerable pecuniary loss. But when the insects have once mounted into the heather twigs and have commenced spinning, there is very little further cause for anxiety, at least so far as that year's produce is concerned.

There are, however, certain imperfections in the cocoons which occasionally occur and render a few of them unfit for commercial purposes. We have already mentioned double and multiple cocoons, that is, such as are produced when two or more silkworms combine to form a single investment for their chrysalises. These, which cannot be reeled in consequence of the interference of the threads, can generally be distinguished by their rounder shape, though occasionally they so closely resemble the normal form as to be practically undistinguishable. It is said that when the worms are allowed to spin as they please in the heather twigs, the proportion of double cocoons is usually from 4 to 10 per cent. of the total.

The perfect cocoon should be hard and tough and capable of resisting pressure at the ends as well as at the sides. The ends should be well rounded, and the thread should become detached easily, with a sort of crackling sound. Finally the colour should not be too bright. In all these respects there are many variations in the cocoons. Sometimes the ends are too thin and yield to pressure; sometimes, again, one end is prolonged to a point and is open at the extremity; such a cocoon would be of no use for reeling, as the water of the reeling basin would gain access to the inside, and thus detach the different patches of silken loops in several parts at once, so that the thread would unwind in lumps. Cocoons with a satiny lustre also are imperfect, the brilliant appearance arising from an incomplete adhesion of the different layers of silk.

If the insect, after having formed its cocoon, should die of an epidemic disease such as flacherie, its body would putrefy and stain the silk black and thus render it valueless. Again, it may die in the chrysalis state, or the newly-hatched moth may be unable completely to extricate itself from the pupa skin. This is the more likely to be the case when the pupa has been removed from the cocoon before the emergence of the moth, which is thus deprived of the leverage the body of the cocoon affords in its efforts to drag itself from the enveloping skin. The struggle thus becomes protracted, and the body, begins to harden before it is entirely freed. Hence the insect is sure to become a cripple, its wings not being duly unfolded ; the same result is produced when the newly disclosed moth has not room enough to expand its wings, and, in fact, any interference at the critical time of its emergence from the pupa case is likely to cripple the insect, in consequence of the softness of all its parts. Cripples should be destroyed, and not allowed to perpetuate their race, even if they are inclined to do so.

By far the most extraordinary of all malformations or monstrosities on record, is a case reported in the " Entomologist" for September 1881, by Mr. E. Kay Robinson. In this instance, the insect only partially emerged from the pupa case, and as much of the contents as was revealed showed *two* moths, a male and a female, in the one pupa, which again originated from a single caterpillar. This is remarkable enough, but the marvel is heightened by the fact that, though the chrysalis itself was of the usual shape, the moths or so much of them as was formed were upside down, that is, their tails were in the position that ought to have been occupied by the head ! But Mr. Robinson had better speak for himself. He says: " The larva, among a small family of six, seemed in no way remarkable, and the pupa was certainly of the ordinary size and shape. In fact, there was nothing noticeable about any of my six silkworms in their immature stages, except that they appeared to get very ill when I fed them on lettuce, and were one and all rather stingy in the matter of silk. Five emerged in the way in which a proper-minded silkworm should, but the sixth, to my astonishment, seemed to have so far forgotten itself as to endeavour to come out tail first from the pupa. My surprise was increased by observing later on that there was evidently another imago of a different sex in the same pupa, equally anxious, and equally unable, to distinguish itself by emerging upside down. From July 14th to 20th they solemnly waved their abdomens in the air without cessation, and then the female tail settled down to the business of life, and laid five eggs, apparently expiring on the 22nd of July. Roused, however, by a stern sense of duty, she revived on July 24th, and laid another egg ; then I think she really died. The male is still—July 26th—alive, though not active ; he seems resigned to his fate of partial imprisonment for the rest of his mortal life. . . . The changes implied in this arrangement are so enormous as to stagger belief, were it not that the specimens are still, and will probably remain, *in statu quo* to prove the matter. It really seems almost pareilel with the case of a man retiring to his room for the purpose of undressing, and then discovering that his clothes contained two persons—a man with his head in one boot, and a woman with her head in the other, and all their feet in his hat!"

Another curious malformation has occasionally been noticed, viz., moths with caterpillar's heads. This results from an imperfeet moulting; the last larva skin remains adhering to the pupa in the region of the head, and is in turn carried off by the moth when it emerges, giving the insect a most remarkable appearance. So perfect is the larval head, that even the tiny eyes are noticeable. Such insects are frequently crippled in the wings.

CHAPTER VI.

WILD SILKWORMS.

WHEN the silkworm epidemics wrought great and continually increasing havoc among the magnaneries of Europe, and seemed to threaten the extinction of the silk industry so far as *Bombyx mori* was concerned, attention was naturally directed to other insects, which could be found in a wild state, and might be fallen back upon as a source of silk supply, in case the ordinary insect should fail altogether. Moreover, as the demand for silk is even now greater than the supply, efforts are still being made to bring under domestication species different from any of the ordinary *mulberry* silkworms, and to introduce silk culture into new districts. These experiments have been made chiefly with Indian and North American insects, and in India alone between fifty and sixty species have been studied with a view to their utilisation. Several of these exotics have been introduced into Europe, and some have been bred even in our own country, so that this little volume would hardly be complete without some notice of what are commonly called "Wild Silkworms," that is, species which, though cultivated, still exist also in the wild state and thus differ from *Bombyx mori*, which is known only under domestication.

Silk producing Lepidoptera belong exclusively to two families, the *Bombycidæ* and the *Saturnidæ*. The former contains only a very few insects, which are of small size and insignificant appearance, and as we have already seen, both the ordinary silkworm

of commerce, and the other mulberry feeding species reared in
India and elsewhere, belong to this family. Notwithstanding
their insignificance as perfect insects, however, and the absence
of any brilliant adornment of colour or grotesqueness of shape in
their larval form, they produce better silk than their more
splendid and remarkable relatives. The Saturnidæ, on the other
hand, are an exceedingly numerous family, containing some four
hundred species already described, and they are noteworthy as
being some of the largest and most magnificent of all moths, and
as possessing brilliantly coloured caterpillars, which are often of
remarkably strange shapes. We shall, in this chapter, be mainly
concerned with the Saturnidæ, which are commonly known as
Emperor Moths. The Bombycidæ, however, demand a few
words first.

Besides the mulberry-feeding species distinct from *B. mori*,
which are domesticated in India and elsewhere, there are also
some truly wild species of the same family, which likewise feed on
different kinds of mulberry and fig trees, and yield silk which is of
some value. Perhaps the most important of these is *B. Huttoni*,
the wild silkworm of the N. W. Himalayas. Its caterpillar,
in colouring and shape, is something like that of the common
silkworm, even to the hump behind the head, but it is covered
with long spines. It is of a pale cream colour mottled with
darker tints. It spins a pale yellow cocoon, which it surrounds
with the leaves of its food-plant, the wild mulberry of the
Himalayan forests, amongst which it occurs abundantly. It is
double brooded, and can only be reared on the trees, being of far
too restless a disposition to remain contentedly in trays filled with
leaves, such as would delight the heart of its civilized, but
captive and enervated cousin.

Captain Hutton, after whom this species is named, remarks on
the instinctive power it seems to possess of detecting the approach
of a hailstorm. For rain it does not care, but hail, of course, is
quite another thing, and the merciless pelting of the icy morsels
would be an extremely uncomfortable, not to say damaging
experience, for such soft-bodied creatures. They do not wait for
the actual commencement of the fusillade, but as the Captain says
of them : " No sooner are peals of thunder heard, than the whole
brood seems to regard them as a warning trumpet call, and all are
instantly in motion, seeking shelter beneath the thicker branches,
and even descending the trunk of the tree to some little distance,
but never proceeding so low down as to lose the protecting shelter
of the boughs."

Besides this insect, there are some seven or eight other wild silkworms which feed on mulberry or fig, some of which yield white and some yellow cocoons. But enough of the unattractive,

Fig. 22.—Tusser Moth (*Antheræa Mylitta*).

yet most valuable Bombycidæ; it is time we passed on to the truly magnificent, though much less used, Saturnidæ. One of the most remarkable peculiarities of these insects in the perfect state

6

is that on each of their four wings they carry a clear spot like a window—an area, usually more or less circular, from which the customary scales are absent both on the upper and under surfaces, only the colourless and transparent membrane, which forms the basis of the wing, being left. These window-like spaces are surrounded by concentric rings of colour, darker than the rest of the wing, and hence the whole appears like a large and conspicuous eye.

By far the most important of these window-winged moths, is a large insect whose silk has been utilised in India and China for many centuries. It is known commercially as the Tusser, Tussah, Tusseh, or Tasar moth, and its scientific name is *Antheræa mylitta* (Fig. 22). It is a noble insect of about six inches in expanse of wings, of a pale creamy yellow in the female, and yellowish brown in the male. A purplish band runs parallel to the outer edges of all four wings, about a quarter of an inch from the margin, and forms an elegant set-off to the uniformity of colour of the main

Fig. 23.—Cocoon of Tusser Moth.

area of the wings ; the eye-spots are very conspicuous, and each clear space is crossed by a fine line. The fore wings are somewhat hooked at the tip, especially in the male, and the antennæ are beautifully fringed.

The caterpillar is a great greenish creature, with rows of reddish warts. It spins a most remarkable cocoon (Fig. 23), of a yellowish grey colour, some darker threads appearing on the outside as a sort of tracery of network. It is perfectly oval, and from near one end there passes a long, hard, and stiff stalk, ending in a loop, which is tightly fastened round a twig. Notwithstanding its hardness and stiffness, this stalk is made of silk, and by its means the cocoon hangs from the twigs of the trees like a fruit. The cocoon is exceedingly tough, and inside it is beautifully smooth—a delightfully snug nest for the dumpy brown chrysalis which has to spend either weeks or months in its seclusion according as it belongs to the early or late brood. So hard and tough are the cocoons, that the natives in some parts of India are said to use them as extinguishers to the bamboo tube in which they keep the tinder used by them for lighting their tobacco and for other purposes. The hardening of the cocoon is produced by the caterpillar discharging upon the newly spun silk a kind of gum or cement which, on drying, binds the silk into a firm, solid mass. The insect escapes from its cocoon in the same way

Fig. 24.—Chinese Oak Silk Moth (*Antheræa Pernyi*).

as the common silkworm, bursting out at the end to which the stalk is attached. The moths do not live long; they soon deposit their eggs, which are large, pale, flattened objects, with two dark reddish rings round the largest circumference.

In their native forests, these insects are very plentiful; they are found over almost the whole of India, and feed on several of the trees which grow luxuriantly there. They have been reared in

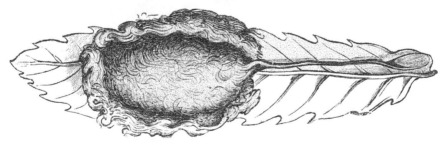

Fig. 25.—Cocoon of *Antheræa Pernyi*.

England by feeding them on the evergreen oak. The life of the caterpillar lasts for about six weeks.

Tusser silk is much coarser than that of the mulberry species, and the fibres are flat instead of round; the width of the thread is about three times that of *B. mori*. It is also very strong, and garments made of it are consequently exceedingly durable. Its colour is pale drab, and it is not nearly so glossy as that produced by the common silkworm, at least when reeled by the natives in India. It is possible, however, by using more care in its preparation greatly to increase the gloss.

Another Indian insect, much like the Tusser Moth, and about the same size, is called the Moonga, or Muga, and scientifically *Antherœa Assama*, because Assam is its chief home. It forms a large, pale yellowish-brown, somewhat oval cocoon, which has not hard walls like the Tusser, but is yielding and flexible. It feeds on a variety of trees, and no less than five broods make

Fig. 26.—Japanese Silk Moth (*Antherœa Yama-mai*).

their appearance in the course of the year. The silk is most brilliant, and only about half the thickness of Tusser silk. The rearing of this insect in India ranks next in importance to that of the Tusser.

A smaller insect, with much less conspicuous eyes on the wings than the Tusser, though similar in colour, is found in Mantchouria, the north-east district of the Chinese Empire. It is the Chinese oak-silkworm, *Antherœa Pernyi* (Fig. 24), and as the common name implies, it feeds on oak. Being an inhabitant of more temperate regions than the preceding moths, it is much more easily acclimatised in Europe, and accordingly has not unfrequently been bred in this country. Mons. A. Wailly, who has had great experience in rearing silk-producing Bombyces in

England, says that this species is very easy to rear on oak trees in the open air, and the moths, too, have been found to pair freely in confinement, a result which is often difficult, if not im-

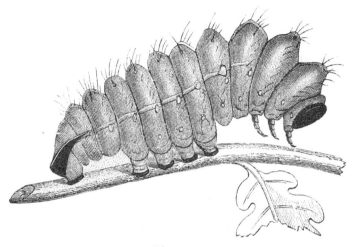

Fig. 27.—Larva of *Antheræa Yama-mai.*

possible, of attainment in the case of exotic species. The cocoon is almost oval, hard, and surrounded with a little loose silk (Fig. 25). It is spun up amongst the leaves of the oak, but, like the Tusser, it has also a long stalk, to assist in mooring it. It is yellowish and glossy. This species was exhibited in the Western world for the first time in 1855.

Very similar to the above insect is *Antheræa Yama-mai* (Fig. 26), the oak-feeding silkworm of Japan. This is a larger species than

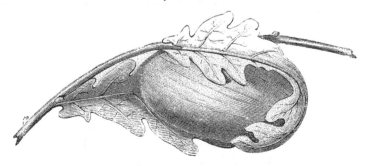

Fig. 28.—Cocoon of *Antheræa Yama-mai.*

the Chinese oak-feeder, and the customary line parallel to the outer margin of the wings is of a blackish hue, giving the moth the appearance of being in half-mourning. The caterpillar is

an oddly-shaped, greenish creature (Fig. 27), and the cocoon
(Fig. 28) is very lovely, being of a delicate pale green colour.
The thread is said to be between two thousand and three thou-
sand feet long.

This species was for a time very jealously guarded by the
Japanese, and it was not till the year 1862 that eggs could be
obtained for transport to Europe.   Strange to say, the introduction
of this Japanese species into Europe in this, the nineteenth
century, is associated with a romantic incident which is the
counterpart of the trick by which the eggs of the Chinese mulberry feeder were conveyed thither thirteen centuries before.   The
eggs were obtained for transportation to France through the efforts
of M. Pompe van Meedervoort, a medical officer in the Dutch
navy, and Director of the Imperial School of Medicine at Nagasaki.   His history of the matter is sufficiently remarkable to
justify us in giving it in its entirety.   He says : " In 1862 I had
the honour to make the acquaintance of M. Eugène Simon.   He
informed me of the great value of the *Bombyx Yama-mai*, and together we made every effort, but in vain, to procure eggs of this
species ; we were told it was absolutely impossible to obtain them.
M. Simon being obliged to return, I made him a promise before
he left to continue my efforts, and, in case of success, to offer the
eggs to the French Government.   But the more I tried, the more
I saw how difficult, if not impossible, was the attempt.   I applied
in vain to the Japanese merchants, the silk-growers, to many
native naturalists with whom I was on friendly terms, lastly, to
the Government, but all in vain.   The reply was, ' The penalty
of death is inflicted on any one who may export these eggs.'
Another idea then possessed me ; to apply to one of my pupils.
As the Principal of the Imperial School of Medicine at Nagasaki,
I was surrounded with students from the different provinces of
Japan, and amongst others from the provinces of Echizen and
Vigo, or Hiogo, where alone the *Yama-mai* silkworms are reared.
One of these youths, who had on several occasions given me
proofs of his extraordinary devotion, was selected by me for the
purpose ; to him I explained the whole affair, and proposed that
he should go to Vigo at my expense, in order to collect and send
me as many eggs as possible.   This brave young man, whose
name I have promised never to divulge, started on the morrow,
and after an absence of fifteen days secretly sent me the eggs,
which he had gathered with much difficulty and danger to himself.   He told me that no one suspected the object of his journey ;
that was in October 1862.   My mission to Japan was finished
November 1st, 1862.   I started for Europe by the English mail

packet, and undertook the charge of carrying these eggs to Europe. This was by no means an easy matter on board a steamship in the Tropics. If the eggs were kept in the cabin, a great risk of their premature hatching was incurred, for the temperature there in the month of November is above 95° F., and in the Red Sea 105° and more. I followed the advice of M. Simon, and placed the eggs in an ice-box on board ship, though often but little ice was therein. To this precaution is due, in a great measure, their safe arrival in Europe in good condition. I arrived at the Hague early in January 1863, and at once sent out the eggs. The greater part were sent to the French Government and to the Imperial Society of Acclimatisation, according to the promise I had made to my friend M. Simon." In this way the insect was introduced into Europe, and since then it has been cultivated in several parts.

The moths appear in August, and soon lay their eggs, which do not hatch till the following April or May ; as the hatching depends on temperature, of course in warmer climates it takes place earlier. The caterpillar, like those of the rest of the genus, is a big green creature. Mr. Wailly gives the following directions for rearing this and other species of silk-producing Bombyces. For the newly hatched larvæ, obtain large bell glasses, having some open ings in the dome, for ventilation's sake, and invert these over saucers full of sand covered with a piece of paper. The food is supplied, not as separate leaves, as we should to the ordinary silk-worm, but in the form of small branches, and these are stuck through the paper into the sand, which serves to keep them fresh. Of course the glasses must not be placed in the sun. One advantage of the paper is that the droppings can be removed by simply blowing them off. When the caterpillars are old enough, they may be transferred to larger branches placed in bottles or jars of water, in such a way that they may not fall in and be drowned. Here they may be left till they form their cocoons, fresh food being supplied as occasion may require. If they have a sufficient supply of nice fresh food, they will not show much inclination to wander. The branches used when they are plunged into the water must be long, else the leaves will become too watery, and injure the larvæ : they should also be cut from the trees in the evening, and not when the sun is shining on them. By June the larvæ may be placed in the open air, in shady and not too dry places.

Three species of oak-feeding silkworms have already been enumerated, but there is yet a fourth, called *Telea Polyphemus* (Fig 29). For this insect we must travel to the United States.

It is very similar in appearance to those already mentioned, but

Fig. 29.—American Silk Moth (*Telea Polyphemus*).

darker. The caterpillar (Fig. 30) is green, striped with white, but when fully grown, it is adorned with a number of silvery and golden spots, which glitter most brilliantly when the sun shines upon it. This insect has been reared in America on an extensive scale. One experimenter who had devoted five acres of young oak trees to the cultivation reports that on his rearing-ground,

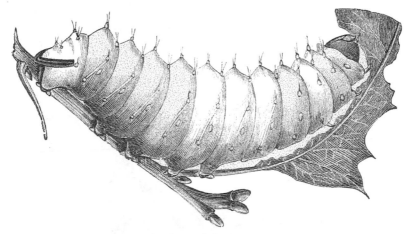

Fig. 30.—Caterpillar of *Telea Polyphemus*.

"not less than one million could be seen feeding in the open air upon bushes covered with a net." It has also been reared in England. The moths appear in May and June, and proceed to lay their eggs on the under side of the leaves. The cocoon (Fig. 31) is not formed till late in September, so that the insect, unlike the common silkworm, passes the winter in the pupa instead of the egg state. In the figure it is shown enveloped in the

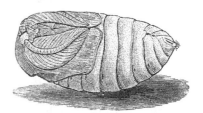

Fig. 32.—Pupa of *Telea Polyphemus*.

Fig. 31.—Cocoon of *Telea Polyphemus*.

leaves of the food-plant. The chrysalis is shown at Fig. 32; its most noteworthy features are its dumpy shape, and the large size of the antennæ cases, the usual characteristics of this group of moths. It thrives well, not only on oak, but on several other trees also, such as nut, birch, and willow. It produces a valuable white silk.

There is a strong family likeness between the species hitherto enumerated. But the genus Attacus, to which we now proceed, represents a different type of insect, in which the prevailing colours are varying shades of red and brown, instead of yellow, and the "windows" are either angular or moon-shaped, instead of circular. The wings, too, are differently shaped; the anterior pair are very decidedly hooked, and the posterior project far beyond the body, which is very short, the two margins of each hind wing being inclined to one another below at an acute angle.

By far the grandest of the members of this section is the gigantic *Attacus Atlas* (Fig. 33) which is noted as being, not only

Fig. 33.—*Attacus Atlas* (much reduced).

the largest of all the Lepidoptera, but if we reckon by the superficial area it covers the largest even of all insects whatever. In the figure its dimensions are reduced to rather less than half the natural size; so that the space really covered by its wings is about four times what is shown in the figure. This truly magnificent creature, which is called in France the "Giant of Moths," inhabits India and China and other parts of South-eastern Asia. It is a familiar object in our own country, as forming usually the centre piece in those small glazed cases of eastern insects one so often sees on sale in the shops of curiosity-mongers. It rests with its wings fully spread, and in this position stretches from tip to tip no less a distance than from seven to ten inches, while occasionally the enormous width of nearly twelve inches is attained. These huge wings, after the fashion of its race, the insect flaps, as it clings to some support, regularly up and down like fans. The "windows" are of enormous size, even for so large an insect; they are triangular in shape, perfectly transparent, and edged with a streak of dark brownish black. As regards colour, the wings are divided into three areas which are ornamented with various shades of reddish brown ; the central division, which contains the "windows," is outlined with dark, succeeding which comes a pale band which on the outer side shades off into a broad dark cloud. An extremely elegant dark wavy line also runs parallel to the outer margins of the wings, and puts a neat finish to the variegated surface.

This splendid insect, which had long been known in the Western world in the dead condition, was seen for the first time alive in Europe in the year 1868, when a number of cocoons sent from India, by Captain Hutton, were hatched in France. It was apparently not till ten years afterwards that a living specimen was first seen in our own country; in the year 1878, M. Wailly had the pleasure of seeing some emerge in London from cocoons he had received from India. Since then it has been bred in this country, the eggs having been laid and all the stages of the insect passed through in the land of its adoption.

Notwithstanding the great size of the perfect insect, the egg is not extraordinarily large, and in fact is considerably smaller than that of the Tusser moth. It is whitish, tinged with reddish brown, an appearance caused by a fluid with which it is wet when laid. The colour is, therefore, not permanent, but can be washed off by immersing in water for a short time. The newly hatched caterpillar is black and white, but, as is so frequently the case with the larvæ of the Lepidoptera, the colours change more or less at each moult, till finally a delicate pea-green hue is acquired; but this colour is modified by a quantity of a white powder consisting of a waxy material, with which the body is covered. There are a number of spines all down the body, which point backwards. The cocoon is enveloped in the leaves of the food-plant, and is irregular in outline, being shaped like a bag; unlike those hitherto referred to, it is not a closed cocoon, but is naturally open at one end, so that no change is produced by the escape of the moth. From the side of the open end, there passes a long silken cord, the further end of which is fastened round a twig, and at the opposite end of the cocoon there is a web-like expansion of silk which helps to fix it. It is of a light brown colour and from two to three inches in length. The chrysalis is bright brown, and for so enormous an insect, is very small. This will not surprise us, however, if we remember that the size of the perfect insect results mainly from the great development of the wings, and that these in

the pupa are in an extremely contracted condition. The silk is
strong and good, but difficult to reel. The caterpillar feeds on a
great number of trees in its native forests, but in Europe it has
been reared chiefly on sallow, barberry, plum, and apple. Specimens of the cocoon and moth of this and other silk-producing
exotics may be seen at the proper season in the Insectarium at the
Zoological Gardens, London.

A much smaller and less richly coloured, though similarly
shaped insect is *Attacus ricini*, the Eria or Arindi silkworm of
North-eastern India. It feeds commonly on the castor-oil plant,
though, like Atlas, it will also eat the leaves of several other
trees. The "windows" in the insect are as great a contrast to
those of the Atlas moth as could well be imagined; they are
reduced to mere crescentic streaks. The wings are prettily
variegated with olive brown, pink, white, and black; the body
carries some dainty tufts of snowy white, and the antennæ are
most deeply fringed. In India the moth lays her eggs round a
twig, and these twigs are sold in the market covered with the eggs,
and frequently with the dead moths adhering to the twig, in the
position in which they expired after their maternal duties were
safely accomplished. The cocoon is very small, less even than
that of *B. mori*, though the moth is so much larger, and the silk is
difficult to reel. The insect, however, is easily reared, and several
broods are produced in the course of the year. The silk, again,
is very strong, and can be manufactured into exceedingly durable
fabrics : indeed, it has been said that the life of one person is
seldom sufficient to wear out a garment made of Eria silk, so that
the same piece descends from mother to daughter. For these
reasons, therefore, although the cocoon is so small, *A. ricini* is a
valuable insect.

There is another insect, very much like the Eria Moth, which
feeds on the False Japan Varnish tree (*Ailanthus glandulosa*), and
is known in consequence as the Ailanthus Moth (*Attacus Cynthia*).
This insect is a native of China. A cross between it and the Eria
Moth has been a good deal cultivated in Europe under the name
of the Ailanthus Moth. It was reared originally in France thirty
years ago, and was thence introduced into other parts, and
amongst them, into our own country, where it has been reared
extensively by Lady Dorothy Nevill at Petersfield, and by Dr.
Wallace at Colchester. The Ailanthus tree had been grown in
Europe almost a century before the insect was introduced, so
there was no difficulty as to the supply of food. The rearing of
this insect is commonly known as Ailanthiculture, and its silk is
sometimes called ailantine.

The late Mr. Frank Buckland thus describes a visit he paid to the ailanthery of Lady Dorothy Nevill: "Her ladyship has set apart a portion of her beautiful and well-wooded garden, and has planted it with young Ailanthus trees, covering them over with a light canvas-made building: a precaution rendered necessary by the birds, who pick off the young worms. On entering the building I saw, for the first time, the living worms; they were in the highest state of perfection, and really beautiful things to look at: not white-faced, pale-looking things like the common silkworm, but magnificent fellows from two and a half to three inches long, of an intense emerald green colour, with tubercles tipped with a gorgeous marine blue. Her ladyship pointed out to me how the silkworms held on to the leaves; they cared nothing for the rain, less for the wind; their feet have greater adhesive power than the suckers of the cuttle-fish, and their bodies are covered with a fine down [rather powder] which turns the rain drops like the tiny hairs on the leaf of the cabbage. Many of them had made their cocoons, picking out snug, quiet corners, and were working away like diligent and useful weavers as they are."

Dr. Wallace's plantation was on the railway embankment between Colchester and Wivenhoe. Here, in 1864, he planted 3,000 trees, two-year-old seedlings which he had obtained from France. By the next year 1,340 of these had attained a sufficient size and luxuriance to be used for feeding purposes; the insects were placed upon them and, as the result of the first brood, a crop 5,368 cocoons was obtained, an average of about four cocoons to a tree. But there are many drawbacks to rearing insects in the open air in this way. Dr. Wallace thus refers to one difficulty which caused him some perplexity for a time: "I was much vexed to find some of my finest larvæ, when nearly full fed, mutilated in a very strange manner; they were resting on a leaf as usual, but with head erect and face looking black. On closer inspection, I found the face eaten away, and nearly gone, and a brownish ichor exuding from the wound; in a few days they died, starved, being unable to eat. I was puzzled for some days; but observing that this always took place at night, and that I never found any larvæ so attacked in the day-time, I suspected a new enemy other than I knew of. Taking my lantern, I sallied out when it was dark, to make a close examination; I found the common large garden *Carabus violaceus* [one of the ground beetles] fastened to the face of a larva, sucking its juices: whether it selected that part for attack I know not, but I am inclined to think, from the habit of the larva to turn round and face any object which touches it, after the manner of the larva of

the puss-moth, and even to open its jaws as if to bite, that when
the Carabus approached the larva, it presented its face as if to
attack, and was then pinned by the Carabus, who, when he got
his jaws in, would soon suck the juices which exude very freely."
Besides these beetles, he found that birds, such as tomtits, thrushes,
robins, rooks, sparrows, magpies, starlings, and jackdaws, did a

Fig. 34.—Eggs, larvæ, and cocoons of Ailanthus Moth.

good deal of damage, so that altogether ailanthiculture in the open
air is rather an anxious business.   Fig. 34 shows the eggs, caterpillars, and cocoons of the Ailanthus Moth.   One cocoon is
enclosed in a leaf of Ailanthus, on which tree the larvæ are feeding,
and the other is unenclosed.

In the breeding of such insects as this, it must be remembered
that, as cripples seldom pair, and are therefore useless for pro--
viding eggs, every facility should be afforded to the moths as they
issue from the cocoons, for the proper drying and expanding of

their wings. To secure this end, it is often the practice to thread the brood cocoons on a string and hang them up in some warm place as the time for the appearance of the moths approaches. In this way, each moth has the cocoon immediately above to help itself out with, and on emergence, is at once in the perpendicular position. A chaplet of this kind, with a number of newly expanded moths upon it, is a very pretty sight. Dr. Wallace furnishes a vivid description of the emergence of a Cynthia from one of the cocoons in a chaplet, which, as it gives a very clear idea of the usual method of such emergences, is worth quoting. He says : " See ! a quiver and a shake runs through that chaplet ! Which is the cocoon wherein newly-born life struggles to be free ? Another and another struggle, and we detect the individual cocoon : see ! a greyish yellow face emerges, then the head, then a leg, then the opposite one ; now a third leg, then a wing on the same side is partially pulled out, the shoulders being bent over on the opposite side ; then a fourth leg (the first two pairs), and now, bending first to one side and then another, the limp wings are withdrawn from the cocoon, and then with the last pair of legs, the large, lax abdomen tumbles over the side, and, as it were, drags the willing insect down to the base of the cocoon, where, firmly fixing its pretty, tiny feet, the moth rests to gather strength for the next task, that of expanding and drying its wings. The whole process of birth does not occupy a minute."

This insect will eat the young shoots of laburnum, lilac, or cherry, if it cannot get its natural food, the Ailanthus tree ; Dr. Wallace also reared a peculiar dwarf form from caterpillars fed on celery. It is an easy insect to rear in the open air, and indeed seems to be the only Asiatic species introduced into Europe that is of a sufficiently hardy nature to become thoroughly acclimatised. The cocoon is small, and open at one end, and therefore, though the silk is strong, its reeling is difficult, hence ailanthiculture is yet little more than a hobby.

Closely related to the above Eastern insects are two from the United States, *Platysamia Cecropia,* and *Callosamia Promethea.* The wings are less deep and angular than those of Cynthia and Atlas, and are coloured with various shades of brown, somewhat relieved with pink. The " eyes " in Cecropia are crescentic, and coloured with pink and white, scarcely a trace of the transparent " windows " being left. The dark brown basal half of the wing is bordered by a band of first white, and then pink, which shades off into brown. The body, which is short, is very prettily banded, the edge of each segment being deeply bordered with a pale fringe. There is also a distinct dark eye-spot near the tip of the

fore wings. Promethea is somewhat similarly coloured, but the
"eyes" are reduced to pale v-shaped marks of which there are
two on each fore wing, and one on each hind wing. The male is
very deeply coloured, but the female much paler and more tinged
with pink.

P. Cecropia is the largest of the American silk producers,

Fig. 35.—Actias Luna.

measuring upwards of six inches in expanse of wings. It feeds on
many trees, such as plum, apple, pear, willow, poplar, maple, etc.,
but it is difficult to rear in this country. It forms an extremely
large brown cocoon, which is open at one end. The caterpillar
of *C. Promethea*, when young, is white, marked with black rings
or spots, but when fully grown is greenish, with black spots, those
on the second and third segments carrying tall, red, cylindrical
prominences. On the last segment but two there is also a central prominence of an orange colour. It feeds on lilac and wild

cherry. The cocoon is only a small one, much like that of Cynthia, and is enfolded in the leaves of the food-plant.

A type utterly unlike anything we have hitherto noticed is to be found in the genus *Actias*. For delicacy of tint and beauty of form, these insects certainly carry off the palm. They are entirely of the most lovely pale green, and the hind wings are provided with a long tail, which somehow or other, in a Lepidopterous insect, always seems to give an air of refinement and aristocratic breeding. As a relief to the pale green ground colour, and in exquisite contrast to it, is a streak along the front edge of the fore wings and running right across the thorax as well, which commencing as black on the outer edge, passes by insensible gradations through shades of red to pure white, where it adjoins the general ground colour. The "eyes" are in the form of a dark crescent on the side nearest the base of the wing, with the rest of the circle filled in with whitish.

Actias Selene is a native of India, and has been reared in this country on walnut, nut, wild cherry, wild pear, etc. There is a very similar but smaller species found in the United States, and called *Actias Luna* (Fig. 35); the specific names of both insects have reference to the crescentric marks which are the most conspicuous adornments of the wings, and strike the observer at first sight; "selene" being the Greek, as "luna" is the Latin, for the "moon." The caterpillar of Selene is an odd-looking creature with the dividing lines of the segments deeply impressed. When young it is dark red, but afterwards green, with a pale streak at the side, and with a number of red warts. It forms a large, rather flexible, dark brown cocoon. The American species feeds on walnut, nut, and other trees, and is easy to rear, but its cocoon is so thin, and the fibre so weak, that it is not likely to be of any use as a silk producer. The caterpillar (Fig. 36), unlike that of its Indian relative, is green all its life.

There is yet another type of these silk-producing *Saturnidæ* that we must mention. It is represented by an Indian species called *Cricula trifenestrata*. This is not a large insect, and in all its stages it can easily be distinguished from any of the foregoing. The moth is brownish yellow, with nothing peculiar about the shape of the wings save that the fore pair of the male are much hooked at the tip. But the distinctive feature is that, instead of having, as so many others of the family have, *one* large "window" on each wing, it carries a row of *three* tiny ones on each fore wing, and a single minute one on each hind wing. It is from this peculiarity that it has derived the name "trifenestrata," or "three-windowed." The caterpillar is rather hairy, and the cocoon is a

7

most extraordinary one. The silk of which it is composed is brilliantly golden and glossy ; it is surrounded by a little " floss," but the "pod," as the firmer part is called, is perforated with large holes, *i.e.*, the silken threads are so arranged in their crossings and interlacings that, instead of forming an entire, compact wall, they leave numbers of openings into the interior, scattered in a perfectly indiscriminate manner over the whole of the cocoon except the point of its attachment to its support. This insect is extremely abundant in the eastern districts of India, where, it is said, the cocoons rot in the jungles for want of gathering. The

Fig. 36.—Larva of Actias Luna.

silk is strong, rich, and glossy, but its cultivation has not yet been developed.

All the insects above enumerated, together with several others which are equally silk-producers, have been at one time or other, bred in this country, and several of them have been reared from one generation to another. The breeders have, however, depended chiefly upon the supply of living cocoons from abroad, from which they rear the moths, and then, if they can secure pairings between these, so as to obtain fertile eggs, they manage with care to conduct the insects through a complete cycle of their meta-

morphoses. But there are many difficulties in the way of success.
Our climate is so variable, that while one season may be
exceptionally fine and therefore yield most encouraging success,
the next may turn out just as wet, and prove in consequence,
most disastrous. In the transmission of cocoons to this country,
especially from the east of Asia, great risks to the life of the
insect are run, and large numbers perish in one way or other on
the journey, so that, unless an abundant supply can be secured,
success is likely to be but trifling.

A few figures from the experience of M. Wailly will well
illustrate this. At the end of February 1880 he received from
Calcutta nine hundred cocoons of the Tusser Moth ; on their
arrival, he found that seven hundred and fifty of them had died
during the passage; out of the nine hundred, therefore, only
a hundred and fifty remained for him to rest his hopes upon.
In the middle of April, in the same year, he received another
consignment of only a hundred cocoons, packed in a tin box.
A totally different disaster had happened to these : the season
being now so much more advanced, and the heat having been
great on the journey, the natural time of hatching had been
anticipated, and more than two-thirds of the moths had emerged
during the voyage, and in the confinement of the box had of
course ruined themselves. A similar fate befell one hundred and
eleven cocoons of the gigantic Atlas Moth received from India on
another occasion. On opening the box, the consignee was
chagrined to find it full of broken wings and dust ; the great
creatures had made their exit from the cocoons during transmission, and had knocked themselves to pieces.

And even if they do travel safely, reach their destination alive,
and issue in due course as moths, it by no means follows that
eggs will be obtained such as shall be suitable for propagating
the race. The moths may issue at irregular intervals, either
singly, or with perhaps several of one sex only at a time, so that,
as they rarely live many days after acquiring their adult form,
those first out may die before there is any chance for them to
get mated. For example, M. Wailly had once sixty cocoons of
A. Selene from which he only obtained *one* pairing, because the
moths emerged at intervals from March to August, and when a
number *did* happen to issue near together, they perversely turned
out to be all of the same sex.

The time of issue of the moths, too, is very irregular when they
are removed from their native climate, and M. Wailly has found
that "tropical species are apt to emerge during the winter when
the weather is mild, while moths of native, or northern foreign

countries seldom, if ever, emerge before spring." Of course, if the eggs are laid at unsuitable times, the caterpillars are apt to hatch when they are not wanted, and when there is not sufficient food for them. Then again, if they are reared in the open air, they are always liable to the attacks of birds, unless protected with nets, and the older the caterpillars are, and the fatter and more luscious they become, the greater sinners do the feathered tribes become against them ; M. Wailly found London sparrows terrible thieves in this respect. Again, after a long continuance of fine, glorious summer weather, during which the caterpillars have thriven well, and have begun to look charmingly prosperous, and to raise the hopes of their cultivator proportionately, there may, in this unreliable climate of ours, come suddenly a change of weather, in the form of drenching rain, or piercing cold, which will speedily blast all hopes of a grand result from the season's care and watching. Birds are, as before intimated, not the only depredators that have to be feared ; M. Wailly once had a quantity of young caterpillars of the Ailanthus Moth destroyed by lady-birds ; he also met with some difficulties on the score of cats. A number of fine larvæ reared indoors, were again, on one occasion, carried off by mice. One gentleman in London has been very successful in rearing these exotic species in a large glass case intended for fern-growing ; this was warmed and kept at a high temperature by means of a paraffin stove, over which was placed a saucer of water to produce the requisite moisture in the atmo-sphere. Notwithstanding the smell of petroleum the larvæ throve well, and seventeen larvæ even of the Atlas Moth, a difficult moth to rear, fed up in less than a month, and formed cocoons from which the moths issued a fortnight after.

It is a curious fact that crosses between several of the species reared have been obtained, and that in some cases, unlike the generality of hybrids they have been able to perpetuate their race, and to continue doing so for several generations. The experi-ments in this direction, however, have yet advanced but little beyond their infancy.

There seems to be no chance of making the rearing of these exotic species a financial success, at least in this country. Still there is a good deal of interest attaching to them, both on account of their large size and of their beauty, not only as moths, but also as caterpillars, and as a mere amusement, the rearing of such splendid insects is likely to prove quite as attractive as that of the unadorned caterpillars and moths of *Bombyx mori*, the chief drawbacks being the amount of care required, and the small proportion of success that must be looked for.

PART II

𝕹atural 𝕳istory and 𝕾cientific 𝕭ooks

PUBLISHED BY

SWAN SONNENSCHEIN & CO.

Elementary Text-Book of Zoology :

By PROF. W. CLAUS (University, Vienna), edited by ADAM SEDGWICK, M.A., Fellow and Lecturer of Trinity College, Cambridge, assisted by F, G. HEATHCOTE, B.A.
With 706 new woodcuts. 2 vols. Demy 8vo, 37*s*., or, separately :

I. General Introduction, and Protozoa to Insecta, 21*s*.
II. Mollusca to Man, 16*s*.

"It is thoroughly trustworthy and serviceable, and is very well got up. The 706 beautifully clear and most judiciously selected woodcuts enhance the value of the book incalculably, and there can be little doubt that it will be universally adopted as an elementary text-book."—*Athenæum.*

"Teachers and students alike have been anxiously waiting for its appearance. . . . We would lay especial weight on the illustrations of this work for two reasons ; firstly, because correct figures are of enormous assistance to the student, . . . and secondly . . . it contains as rich a supply of well-drawn, well-engraved, and well-selected figures as ever man could desire. Admirably printed. . . . The whole enterprise reflects the greatest credit."—*Zoologist.*

"It is not often a work so entirely fulfils its object. . . . It is alike creditable to author, translators, and publishers, who seem to have vied with each other in rendering it not only valuable but attractive."—*Knowledge.*

"The exhaustively minute and well-arranged treatment, aided by diagrams and illustrations of wonderful clearness, at once command for this book its proper place as our leading text-book of zoology."—*Glasgow Herald.*

A Treatise on Animal Biology :

By ADAM SEDGWICK, M.A., Fellow and Lecturer of Trinity College, Cambridge. [*In preparation.*

Handbook of Entomology. :

By W. F. KIRBY, of the British Museum.
Illustrated with several hundred figures.
Large square 8vo, cloth gilt, gilt top, 15*s*.

"It is, in fact, a succinct encyclopædia of the subject. Plain and perspicuous in language, and profusely illustrated, the insect must be a rare one indeed whose genus—and perhaps even whose species—the reader fails to determine without difficulty. . . . The woodcuts are so admirable as almost to cheat the eye, familiar with the objects presented, into the belief that it is gazing upon the colours which it knows so well. . . . Advanced entomologists will obtain Mr. Kirby's fine volume as a handy book of reference ; the student will buy it as an excellent introduction to the science and as an absolutely trustworthy text-book.'—*Knowledge.*

The Insect Hunter's Companion:

By REV. JOSEPH GREENE, M.A.
Third edition. With cuts. 12mo, boards, 1s.

Our Summer Migrants:

An Account of the Migratory Birds which pass the Summer in the
British Islands.

By J. E. HARTING, F.L.S., F.Z.S.,
Author of "A Handbook of British Birds," a new edition of White's
"Selborne," etc., etc.

Illustrated with 30 Illustrations on Wood, from Designs by THOMAS
BEWICK. 8vo, cloth elegant, 7s. 6d. [*Second Edition shortly.*

Dictionary of British Birds:

By COLONEL MONTAGUE.
New edition. Edited by E. NEWMAN, F.L.S.
Demy 8vo, cloth gilt, 7s. 6d.

Brown's Conchology:

A new edition, brought down to date, with all the original copper
plates, is in preparation, and will be issued shortly. Royal 4to, with
several thousand coloured figures.

Elementary Text-Book of Botany:

By PROF. W. PRANTL and S. H. VINES, D.Sc., M.A., Fellow and
Lecturer of Christ's College, Cambridge.

Fifth edition (1886). Illustrated by 275 woodcuts. Demy 8vo,
cloth, 9s.

[This book has been specially written as an Introduction to SACHS'
"Text-Book of Botany," at the request of Professor Sachs himself.]

"It is with a safe conscience that we recommend it as the best book in
the English language."—*Nature.*

Elementary Text-Book of Practical Botany:

A Manual for Students. Uniform with the above.

Edited from the work of PROF. W. STRASBURGER, by PROF. W.
HILLHOUSE, M.A., of the Mason College, Birmingham.

Illustrated by a large number of new woodcuts. 8vo, 9s.

"The chief features of the author's work are the numerous well-selected
types dealt with, and the thoroughness with which they are treated. . . . As
might be expected, there is a masterly chapter on the study of the cell and
nucleus and their division. . . . The work is illustrated with numerous
excellent woodcuts. . . . As an exposition of the new methods of botanical
research, it is the best handbook we have yet seen, and should be at hand in
every laboratory."—*Athenæum.*

Life Histories of Plants:

With an Introduction to the Comparative Study of Plants and Animals
on a Physiological Basis.

By PROF. D. McALPINE.

With over 100 Illustrations and 50 Diagrams.   Royal 16mo, 6s.

" A piece of specialist's work, with all the marks of thoroughness and finish
which should distinguish it."—*St. James's Gazette.* -

Handbook of the Diseases of Plants:

By PROF. D. McALPINE.

Illustrated.   Demy 8vo.   [*In preparation.*

Alpine Plants:

Painted from Nature, by J. SEBOTH, with descriptive text by A. W.
BENNETT, M.A., B.Sc.

4 vols., each with 100 Coloured Plates.   Half Persian, each 25s.

Tourists' Guide to the Flora of the Alps:

Edited from the work of PROF. K. W. v. DALLA TORRE, and issued
under the auspices of the German and Austrian Alpine Club in Vienna.

By A. W. BENNETT, M.A., B.Sc.

Elegantly printed on very thin but opaque paper, 392 pp., bound as
a morocco pocket-book, pocket size, 5s.

" This handy little volume looks like a pocket-book, as it should, and
constitutes a sort of botanist's *vade-mecum*, affording short descriptions of
most of the Alpine plants so dear to visitors. . . . All the most interesting flowers of the Alps are here. . . . The classification adopted is the
one most commonly met with in English floras, and the natural orders are
old friends. . . . It is excellent for its purpose, and reflects great credit
on all concerned."—*Athenæum.*

A History of British Ferns:

By E. NEWMAN, F.L.S.

Third Edition. ' Cuts.   Demy 8vo, cloth, 18s.

A "People's Edition" of the above (abridged), containing
numerous lithographed figures.   Fifth Edition.   12mo, cloth, 2s.

An Elementary Text-Book of British Fungi:

By W. DE LISLE HAY, F.R.G.S.

With about 400 large Figures.   Royal 8vo, cloth extra, gilt top, 15s.

" A useful and trustworthy manual."—*Scotsman.*

The Fungus Hunter's Guide:

By W. DE LISLE HAY, F.R.G.S.

Fully Illustrated and interleaved.   Crown 8vo, limp cloth, 3s. 6d.

11

A Season among the Wild Flowers :

By the REV. H. WOOD.

Second edition. With numerous Cuts. Crown 8vo, cloth gilt, gilt edges, 2s. 6d.

"Perfectly free from misplaced raptures, the book is also attractive from its correctness. The plates are unusually and, indeed, remarkably good for a cheap and popular treatise."—*Academy.*

Flowers and Flower-Lore :

By the REV. HILDERIC FRIEND, F.L.S.

Illustrated. Third edition in 1 vol., 8vo, cloth extra, gilt top, 7s. 6d.

CONTENTS. — The Fairy Garland—From Pixy to Puck—The Virgin's Bower—Bridal Wreaths and Bouquets—Garlands for Heroes and Saints—Traditions about Flowers—Proverbs of Flowers—The Seasons—The Magic Wand—Superstitions about Flowers—Curious Beliefs of Herbalists—Sprigs and Sprays in Heraldry—Plant Names — Language of Flowers—Rustic Flower Names—Peculiar Usages—Witches and their Flower-lore. [*See* p. 37.

"A full study of a very fascinating subject. . . . His two attractive volumes form a perfect treasury of curious and out-of-the-way flower learning. . . . We find also very copious critical and bibliographical notes, with full indices. Altogether the work is an important and exhaustive one, and occupies a distinct place of its own."—*Times.*

Moon Lore :

By the REV. TIMOTHY HARLEY, F.R.A.S.

Illustrated by facsimiles from old prints and scarce wood-blocks. 8vo, cloth extra, gilt top, 7s. 6d.

CONTENTS. —Introduction—The Man in the Moon—The Woman in the Moon—The Hare in the Moon—The Toad in the Moon—Other Moon Myths—Moon Worship—Moon a Male Deity—Moon a World-wide Deity—Moon a Water Deity—Moon Superstitions—Lunar Fancies—Lunar Eclipses—Lunar Influences—Moon Inhabitation. [*See* p. 37.

"A pleasant excursion into one of the by-paths of literature, it brings together a mass of facts, traditions, and notions concerning the moon collected from an infinite variety of sources, and never before included within the covers of a single volume."—*Scotsman.*

"A chatty, charmingly written book on 'Moon Lore.' Mr. Harley is very thorough, as well as very amusing."—*Graphic.*

Lunar Science :

By the same Author. Supplementary to "Moon Lore." 3s. 6d.

"A very popular and readable account of facts known about the Moon.—*Journal of Microscopy.*

"Told in a popular and entertaining manner."—*Bookseller.*

An Elementary Star Atlas :

A Series of Twelve Simple Star Maps, with Descriptive Letterpress, for the use of beginners with the telescope and naked-eye star-gazers. By the REV. T. H. E. C. ESPIN, B.A., F.R.A.S., Special Observer to the Liverpool Astronomical Society.

4to, cloth, 1s. 6d.

"These maps are of a convenient and handy size, and their arrangement is good."—*Athenæum.*

"A tempting popular introduction to the fascinating science of astronomy."—*School Board Chronicle.*

Bibliography, Guide and Index to Climate:

By A. RAMSAY, F.G.S.

With a few Woodcuts. Demy 8vo, cloth gilt, 16s.

"A most valuable addition to every scientific, public, or reference library." —*Saturday Review.*

"It forms a large volume of 500 pages, and tabulates a vast mass of interesting and valuable matter bearing on Meteorology. . . . Great research is exhibited."—*Times.*

Tabular View of Geological Systems:

By DR. E. CLEMENT.

Crown 8vo, limp cloth, 1s.

"Shows at a glance the order of the geological systems, and the divisions and subdivisions, the places of their occurrence, their economic products, and the fossils found in them."—*School Board Chronicle.*

The Naturalist's Diary:

A Day Book of Meteorology, Phenology, and Rural Biology.

By CHARLES ROBERTS, F.R.C.S., L.R.C.P., etc.

With a Coloured Folding Flower Chart. 8vo, limp cloth, 2s. 6d.

The "Naturalist's Diary" is intended to be used as a work of reference on many questions relative to Natural History, Climate, Periodic Phenomena, and Rural Economy ; and as a Journal in which to record new facts and observations of a similar kind.

"A delightful device. . . . Will make every man his own White of Selborne."—*Saturday Review.*

Electricity v. Gas:

By JOHN STENT.

Crown 8vo, paper boards, 1s.

The Microscope:

By PROF. C. NAEGELI and PROF. S. SCHWENDENER.

With a Preface by FRANK CRISP.

With about 300 Woodcuts. Demy 8vo, cloth, 21s.

"Renders an important service to students by reason of the characteristic thoroughness of the exposition, both of theory and practice."—*Daily News.*

Evolution and Natural Theology:

By W. F. KIRBY, of the British Museum.

Crown 8vo, cloth, 4s. 6d.

"A book of much interest from the pen of a ready writer."—*Knowledge.*

"There is a great deal of interesting and curious matter in the volume."—*Science Monthly.*

Simple Mechanics:

A Practical Guide for the Home and the Workshop.

Adapted to the Daily wants of Everybody.

By GEORGE E. BLAKELEE.

With over 200 large Illustrations.

720 pages, royal 8vo, cloth gilt, 15s.

The Dynamo:

How Made and How Used. By S. R. BOTTONE.

Second Edition, with 39 large Illustrations. Crown 8vo, cloth, 2s. 6d.

"Exceedingly plain, clear instructions for the manufacture of small dynamos."—*Journal of Science.*

"Gives minute instructions to amateurs, by following which they may be able to make, at trifling cost, a dynamo for themselves."—*Scotsman.*

The Little Cyclopædia of Common Things:

By SIR GEORGE W. COX, Bart., M.A.

Sixth Edition. Well Illustrated. Demy 8vo, cloth gilt, 7s. 6d.

" Has deservedly reached a third edition. For handy reference and information on subjects of common interest, it is to be preferred to the big encyclopædias. You get an explanation, for example, concerning the raw materials and products of manufacture, the practical applications of science, and the main facts of natural history, chemistry, and most other departments of knowledge, within brief compass. . . . The numerous illustrations are often a material help in clearing away difficulties and misapprehensions that widely prevail with regard to common things. The volume has also the mportant recommendation of being remarkably cheap."—*Scotsman* (1884).

The Wanderings of Plants and Animals:

By PROF. VICTOR HEHN. Edited by J. STEVEN STALLYBRASS. Demy 8vo, cloth extra, gilt top, 16s.

CHAPTERS : The Horse—Vine—Fig-tree—Olive-tree—Tree Culture —Asses, Mules, Goats—Stone Architecture—Beer—Butter—Flax and Hemp—Leek—Mustard—Lentils and Peas—Laurel and Myrtle—Box-tree— Pomegranate— Quince—Rose— Lily—Saffron—Date—Cypress —Plane—Pine—Fowl— Pigeon—Peacock—Pheasant—Goose— Hawk-ing—Cat—Rabbit—Hop—Rice, etc. [*See also page* 36.

The Natural History and Antiquities of Selborne:

By the REV. GILBERT WHITE, M.A.

The Standard Edition by BENNETT.

Thoroughly Revised, with Additional Notes, by JAMES EDMUND HARTING, F.L.S., F.Z.S.,

Author of "A Handbook of British Birds," "The Ornithology of Shakespeare," etc.

Numerous Engravings by THOMAS BEWICK, HARVEY, and others. Thick 8vo, cloth extra, gilt edges, 7s. 6d.

The Farmer's Friends and Foes:

A Popular Treatise on the various Animals which affect British Agriculture beneficially or injuriously.

By THEODORE WOOD.

Fully illustrated. Crown 8vo, cloth 3s. 6d.

[*Just published.*

The Dog:

Its Management and Diseases.
By PROF. J. WOODROFFE HILL.
Illustrated. Revised edition brought up to date.
Demy 8vo, cloth, 7s. 6d. [*Just published.*

The Book of the Cat:

By RALPH O. EDWARDS, F.Z.S.
With notes on Feline Diseases by PROF. J. WOODROFFE HILL.
Fully Illustrated, 4to. [*In the press.*

Poultry:

A Manual for Breeders and Exhibitors.
By a POULTRY FARMER.
Fully illustrated with large Plates, in 4to, cloth gilt, 3s. 6d.
[*Just published.*

Rabbits: for Exhibition, Pleasure, and Market:

By RALPH O. EDWARDS, F.Z.S., assisted by several eminent breeders.
With eight plates. Second edition, enlarged.
Crown 8vo, cloth, 2s. 6d.

Also, **Food for Rabbits.** By the same. Price 1s.

Minor Pets: Their General Management, etc.:

(Guinea Pigs, Rabbits, Fancy Mice, Hares, Squirrels, etc.)
By RALPH O. EDWARDS, F.Z.S. With a Chapter on the DORMOUSE
by W. T. GREENE, M.A., M.D., F.Z.S.
With nine full-page Plates and several Woodcuts.
Crown 8vo, cloth, 2s. 6d.

The Cat:

Its Varieties, Diseases, and Treatment.
By PHILIP M. RULE.
With four plates. Crown 8vo, cloth, 2s. 6d.

Handbook of Agriculture:

By R. EWING.
With a Preface by PROF. JOHN SCOTT.
12mo, limp cloth, 6d.

Ensilage, and its Prospects in English Agriculture:

By PROF. J. E. THOROLD ROGERS, M.P.
Second Edition. Cuts. Crown 8vo, limp cloth, 1s.

Horticultural Buildings:

Their Construction, Heating, Interior Fittings, etc., with Remarks on
the Principles involved, and their application.
By F. A. FAWKES, F.R.H.S.
Numerous Woodcuts. Crown 8vo, cloth, 3s. 6d.

Natural History Handbooks for Collectors:

Each Volume is very fully illustrated with practical woodcuts, and bound in flat cloth extra, 1s. each (post free, 1s. 2d.).

"They contain just the kind and amount of information required. . . . It is not easy to understand how works like these, written by men of science in the various departments, can be made a commercial success. Certainly nothing but the enormous circulation which they well deserve, can render them so."—*Knowledge.*

"We have seen nothing better than this series. It is cheap, concise, and practical."—*Saturday Review.*

"We are glad to call attention to this excellent series of handbooks, which deserve to be widely known. . . . We are glad to see the staff of the British Museum thus coming forward to make popular the stores of learning which they have. . . . The illustrations are uniformly good—far better than in many expensive books."—*Academy.*

1. Butterflies, Moths, and Beetles.
By W. F. KIRBY, of the British Museum.

"A really admirable and absurdly cheap manual. The incipient entomologist will do himself an injustice if he does not procure it."—*Knowledge.*

2. Crustaceans and Spiders.
By F. A. A. SKUSE.

"The descriptions are plain and easy, and will serve to help the young student to know and classify his sample whenever he catches it."—*School Board Chronicle.*

"The illustrations are quite up to the standard of this useful series."—*Saturday Review.*

3. Fungi, Lichens, etc.
By PETER GRAY.

4. Mosses.
By JAMES E. BAGNALL, A.L.S.

"Really a wonderful shilling's worth. It is an excellent introduction to the study of mosses. Any one who knows Mr. Bagnall would naturally expect such a guide. . . . No one can hesitate to order copies."—*Grevillea.*

5. Pond-Life.
By E. A. BUTLER, F.Z.S.

"Sound in exposition and excellent in method, this little book is just the right companion for the young naturalist on his rambles. The woodcuts are numerous and good."—*Saturday Review.*

6. Seaweeds, Shells, and Fossils.
By PETER GRAY and B. B. WOODWARD, of the British Museum

7. Ants, Bees, Wasps, and Dragon-flies.
By W. HARCOURT-BATH.

Extra Series.

Coins and Tokens (English).
By LLEW. JEWITT, F.S.A.

With a Chapter on Greek Coins. By BARCLAY V. HEAD, Brit. Mus. The New Volumes shortly to be issued are: **Reptiles,** by CATHERINE HOPLEY. **British Birds,** by R. BOWDLER SHARP and W. HARCOURT-BATH. **Silkworms,** by E. A. BUTLER. **Land and Fresh Water Shells,** by DR. W. WILLIAMS, Editor of the *Naturalist's Monthly*. **Fishes,** by F. A. SKUSE. **Mammalia,** by F. A. SKUSE. **The Infinitely Little,** by Dr. W. WILLIAMS.